Definitive Refutations of the Einsteinian General Relativity

Edited by
M.W. Evans

Viva Books
New Delhi | Mumbai | Chennai | Kolkata
Bengaluru | Hyderabad | Kochi | Guwahati

First Indian Edition 2013

VIVA BOOKS PRIVATE LIMITED

- 4737/23 Ansari Road, Daryaganj, New Delhi 110 002
  E-mail: vivadelhi@vivagroupindia.net, Tel. 42242200
- 76, Service Industries, Shirvane, Sector 1, Nerul, Navi Mumbai 400 706
  E-mail: vivamumbai@vivagroupindia.net, Tel. 27721273, 27721274
- Jamals Fazal Chambers, 26 Greams Road, Chennai 600 006
  E-mail: vivachennai@vivagroupindia.net, Tel. 28290304, 28294241
- B-103, Jindal Towers, 21/1A/3 Darga Road, Kolkata 700 017
  E-mail: vivakolkata@vivagroupindia.net, Tel. 22836381, 22816713
- 7, Sovereign Park Apartments, 56-58, K. R. Road, Basavanagudi, Bangalore 560 004
  E-mail: vivabangalore@vivagroupindia.net, Tel. 26607409, 26607410
- 101-102 Mughal Marc Apartments, 3-4-637 to 641, Narayanguda, Hyderabad 500 029
  E-mail: vivahyderabad@vivagroupindia.net, Tel. 27564481, 27564482
- Beevi Towers, First Floor, SRM Road, Kaloor, Kochi 682 018, Kerala
  E-mail: vivakochi@vivagroupindia.net, Tel: 0484-2403055, 2403056
- 232 GNB Road, Beside UCO Bank, Silpukhuri, Guwahati 781 003
  E-mail: vivaguwahati@vivagroupindia.net, Tel: 0361-2666386

www.vivagroupindia.com

Published by arrangement with
Cambridge International Science Publishing
7, Meadow Walk, Great Abington
Cambridge CB1 6AZ, UK

ISBN : 978-81-309-2300-0

Published by Vinod Vasishtha for Viva Books Private Limited, 4737/23, Ansari Road, New Delhi-110 002. Printed and bound by Syndicate Binders, New Delhi.

Contents

Meaningless relativity

During the heyday of relativity Einstein published a short book entitled 'The Meaning of Relativity' which tried to explain the subject in words. It is doubtful whether this book helped anyone understand the subject, and there were many subsequent attempts to explain it in words. They just made the dogma foggier. The literature is saturated with bellowing about how Einstein explained the perihelion advance of Mercury and the deflection of light by the sun. As a matter of fact he did neither. As a by product of my research in the past decade I have had occasion to study Einstein's work in great detail, and have found numerous errors in it. Up until nine or ten years ago I had not studied general relativity, being a chemist by training. I did not assume that it was correct, I simply had not studied it. I had read books that were supposed to explain it, the best being by Pais when I was at University of North Carolina, Charlotte. However, I found that explanations in words inhibited understanding so I tended to leave these books aside.

Physics is saturated with dogma, the most rational subject is in fact the most irrational. This type of problem comes from poor historical scholarship and poor understanding of fundamentals. It is always claimed that Newton explained the three Keplerian laws of orbits with an inverse square law called universal gravitation. He did no such thing. My ancestor John Aubrey could have told you that in the seventeenth century. In the online 'Brief Lives', anyone can see that it was Robert Hooke who discovered that law. When the mathematics are carefully worked out, the inverse square law does not explain the existence of orbits at all, because it is a law of attraction. An orbiting object cannot stay in orbit if there is only a force of attraction. There must be a counterbalancing force of repulsion – common sense. In the foggy dogma or fogma this is 'the centrifugal force', which regrettably does not exist in Newtonian dynamics. So Newton did not explain orbits at all, and he did not discover the inverse square law. The strangest thing is that everyone seems to know this and still carry on chanting fogma. That is human nature. Everyone who has ridden a bike or driven a car knows that the outward force is real enough, but it does not appear in Newtonian dynamics. It is in fact due to torsion of spacetime. The 'centrifugal force' as it is labelled does not come from a 'centrifugal potential energy' as it should. The 'centrifugal potential energy' is really the angle dependent part of the kinetic energy of a free particle moving around a plane, the rotational kinetic energy of a free particle. However, the fact that the free particle does not move in a straight line means by Newton's first law that it has to be acted upon by an external force, an external force that must come from a potential energy by definition. However the rotational kinetic energy is part of that of a free particle, not acted upon by an external force. There is a diametric self contradiction - a free particle is not acted on by any force, but a force is needed to keep it rotating in a plane. It is free and at the same time not free. In the Newtonian dynamics of a particle moving in a straight line there can only be translational kinetic energy. The free particle cannot move in an ellipse for example, without a force being present, but this force is not present and is contrived artificially. In the fogma this rotational kinetic energy is incorrectly called a potential energy and that is that, we have a magically 'effective' potential energy. The basic problem is that Newtonian dynamics deals only with straight lines, not with motions of orbits. Coriolis attempted to cure this problem but the explanation is again artificial. The real explanation for orbits is that they are generated by the torsion of spacetime itself, and torsion is something that Einstein never considered. I found out recently that if I just differentiated the functional equation of the ellipse (the observed orbit of a planet to an excellent approximation) the inverse square law for force is obtained, with its negative sign indicating attraction, but there is nothing to counterbalance the attraction. The functional equation of the ellipse is the dependence of the radial vector on the angle, and the radial vector twice differentiated is the force divided by the mass of the planet. So what is always known in the dogma as an inverse square law of attraction is merely a re-expression of the equation of the ellipse. There is nothing to prove that there is an attraction or gravitation, these are anthropomorphic concepts, there is nothing to prove that there is an outward force. All we all left with is an observation, that the orbit is an ellipse. I suspect that this is what Newton really did, he used the techniques of his time to differentiate the ellipse twice to find an inverse square law. Newton's greatest contribution in my opinion is mathematical, the inference of differentiation and integration which he called 'fluxions'. When Hooke first drew Newton's attention to the problem, Newton certainly did not get the right answer, he thought the ellipse was due to an inverse r law. There is no reason to doubt John Aubrey in his classic 'Brief Lives'.

The dogmatic idea of a universal law of gravitational attraction arose from a coincidence described as follows. Using cylindrical polar coordinates in a plane consider the function of an ellipse, r as a function of theta, where r and theta are the coordinates. Differentiate the position vector r twice with respect to time using the correct rules for differentiation in this system of coordinates. The result is a negative valued linear acceleration vector inversely proportional to r squared and directed along the radial unit vector. The proportionality constant between this acceleration and the inverse square of r is made up of constants of the ellipse. These are the right magnitude and the ellipticity. Now apply the rule known as Newton's 'second law', a rule which merely defines the concept of force (which first occurs in Kepler's writings) as being linear acceleration multiplied by a mass m and which is not an inference based on data as in the usual meaning of 'law' in physics. Differentiating the ellipse twice therefore produces a force that is negative valued, directed only along the radial unit vector, and which is inversely proportional to r squared. In so doing the equivalence of inertial and gravitational mass m has been assumed. It has been assumed that m is the same in Newton's second law and 'his' law of attraction first inferred by Hooke from the work of Kepler. If the ellipse is assumed to be the orbit of m around M then Newton's inverse square law is the force needed to maintain this motion indefinitely, or, more accurately, produced by this motion. The radial vector is the line joining m and M. Note carefully that there is only one, negative valued, acceleration and only one force. There is no 'centrifugal' force and no 'centrifugal' acceleration. By experimental observation, the force in the laboratory between a static mass m and static mass M is the same inverse square law with proportionality constant mMG, where G is known as Newton's constant. This is however pure coincidence, if the ellipse is made to precess the force law becomes a sum of terms, one an inverse square law in r, the other an inverse cubed law. For galaxies the force law is different again, and wholly different from an inverse square law. There is no 'universal' force of gravitation.

Even worse for the dogmatist is that the negative sign of the force of attraction between static m and static M in the laboratory is merely a convention. If an apple of mass m drops to the ground the dogma claims that it is due to the negative valued force of attraction exerted by the earth, of mass M. The negative sign of the force obtained by differentiating the ellipse is a mathematical property of the ellipse, more accurately a geometrical property intrinsic to the ellipse itself. The negative sign does not originate in the same source, the two phenomena (elliptical orbit and apple falling to the ground) are different. Therefore the force between m and M of an elliptical orbit is not a force of attraction as in the dogma. If it were, m would fall into M, contrary to observation. The true source of the negative valued acceleration of the ellipse is simply the ellipse and nothing else. The ellipse may be described as a Cartan tetrad and the spacetime torsion due to the ellipse evaluated by the first Cartan structure equation of differential geometry. The origin of the elliptical orbit is one kind of spacetime torsion. The origin of the central force between static m (apple) and static M (earth) is another, entirely different, kind of spacetime torsion. As described already the dogma asserts that an elliptical orbit is a balance of a negative valued force of 'attraction' and a 'centrifugal force of repulsion'. Nether force exists, the former is a property of the elliptical tetrad and its negative sign does not denote 'attraction' - an anthropomorphic concept. The 'centrifugal force of repulsion' is confused in the Newtonian dogma with the rotational kinetic energy. The latter is merely another definition, it comes from the use of the work integral and the Newtonian definition that force is mass multiplied by acceleration.

It is easy to see with a little thought that all kind of confusion has crept into physics since the seventeenth century. It is not surprising the general relativity has also collapsed like a pack of cards. There is no purpose in obliging the student to reiterate dogma known to be incorrect or completely confused. A reform in physics education is long overdue.

So from the very beginning, Einstein set out to 'correct' dogma, Einstein's theory in the solar system is a tiny correction of Newton's theory, which does not explain orbits at all, it is not possible to correct something that does not explain anything. Yet this is Einsteinian general relativity, claimed to explain nothing with ever increasing precision at suitably astronomical expense. Outside the solar system in such objects as the whirlpool galaxy, Einstein's theory does not begin to explain the velocity curve of that whirlpool galaxy. It fails hopelessly, and so does Newton's theory. The dogmatists introduced more useless dogma – dark matter, and dark indeed. In the last decade I discovered that the existence of a whirlpool galaxy is due again to spacetime torsion, which throws the stars outwards in their observed spirals. In the solar system the spacetime torsion throws the planets into a precessing ellipse. There is no gravitation and no force of attraction, which is precisely the conclusion found by differentiating an elliptical orbit of the solar system and just explained. Motions and dynamics are due to torsional motions of spacetime itself.

So how did Einstein's dogma become so entrenched? I have no doubt that there is merit in Einstein's work in relativity, merit in its basic and ancient idea that physics is geometry, but a lot of it went wrong very early on and in order to see this with clarity a little accurate history is needed. The first steps towards the idea that physics is geometry were made by the ancient peoples such as the Greeks, who had the idea that geometry is an expression of perfection and therefore of beauty. The proportions of the Parthenon were worked out according to this idea, and a flaw in the construction was put there so as not to offend the gods. Only the gods could know perfection. The kind of geometry developed by thinkers such as Euclid (a corruption of his real name) is known as geometry in space, for example a two dimensional or three dimensional space. Physics is automatically an expression of geometry because quantities of physics such as vectors are expressions of geometry. Newton wrote his Principia entirely in terms of geometry. This is a kind of relativity in space. Naturally, an object moving forward at a given speed with respect to another means that the second object is moving backward at the same speed with respect to the first. This observation is made whenever a car starts to move backwards when one is sitting in a bus until one realizes that the bus is moving forward and the car is stationary. This kind of relativity is one of those things that are accepted as obvious, until there arrives the need to change the entirely obvious, and this is how great discoveries are made.

It appears to have been Riemann who introduced the idea that the geometry of Euclid could be made into another kind of geometry which reduces to that of Euclid but is at the same time more general. Riemann introduced the metric as a measure of the way in which his geometry differed from that of Euclid, whose metric using static Cartesian coordinates is a diagonal, three by three, unit matrix. In Riemann's geometry the object known as the metric can become different from the unit diagonal, a typical mathematician's idea one may think, devoid of all realism. These mathematical ideas have a habit, though, of becoming physics, which is the philosophy and description of nature or 'natural philosophy'. The greatest advances in physics are made by controlling the mathematics, so that it does not become incomprehensible. If that happens, enlightenment is lost and we are back to dark matter again, or even worse, the Higgs boson. Riemann worked in the early part of the nineteenth century and worked only with the idea of the metric.

In the eighteen sixties Christoffel introduced another way of making Euclid's geometry more general. Christoffel proceeded by developing the idea of the derivative using an object known very obscurely as 'the connection'. Only mathematicians would think of such an utterly obscure label. In the Euclidean geometry the derivative can be worked out with the every day Cartesian coordinates, which define a static frame of reference. If however the cylindrical polar coordinates are used, the axes themselves move in time, and the derivatives are defined with extra terms as a consequence of the movement of the axes themselves. This example encapsulates the idea of the connection, the word comes from the fact that the frame of reference itself is no longer static, the state of the frame at one point in a given mathematical space is connected to the state of the frame at another point. The simplest example is rotation of the frame in a two dimensional space, the connection is then related to a rotation matrix. Almost always, the rotation is thought to be the movement of a vector clockwise for example, with static Cartesian axes. However, the same rotation can be described by keeping the vector constant and moving the axes counterclockwise. This can be described as 'rotational relativity'. The rotation can be defined by a rotation matrix, and this matrix is an example of a connection matrix.

The first piece of confusion I noticed in my purely historical researches in this subject is the mis-attribution to Riemann of the concept of connection, so it is known quite wrongly as 'the Riemann connection'. It is the Christoffel connection, Riemann inferred only the metric, which is a symmetric tensor in contemporary language. Again, Riemann had no concept of 'tensor', this idea was introduced much later by mathematicians including Levi-Civita, Ricci and Bianchi. Riemann had no concept of 'vector', an idea introduced by Heaviside and Gibbs long after Riemann's time. Such confusion did not give me much confidence in the scholarship of those who claim to be experts in general relativity. The confusion about centrifugal force does not give me much confidence in school teachers who continue to befuddle their pupils with dogma, as if the teachers themselves are confused and are just doing a job to earn a salary, or even worse, drifting through a 'career'. The crass mis-attribution of the Hooke law to Newton is very poor scholarship by those who claim to be seeking the truth about nature. These are examples of what I call 'fogma' or foggy dogma coming from a laziness of mind and in the worse case, catatonic boredom. They should have gone into another subject.

In its full glory the Christoffel connection is an object with three indices, usually written with two subscripts and one superscript. This is why it is known as the 'Christ awful' connection – it looks like a hedgehog or porcupine into which no dog (short for dogmatist) would like to put its nose. The necessity for these three indices comes from the need to define a derivative in such a way that that derivative is a tensor known as the covariant derivative. With this definition the connection takes on a life of its own, it is not a tensor: under the most general kind of coordinate transformation it does not retain its mathematical structure – a nasty inhomogeneous term appears and the connection is not generally covariant. At this point most dogmatists give up because they do not go further into the mathematics as it starts to get difficult, and the dogmatist becomes just that, a person who relies endlessly on the work of someone else, and repeats someone else's ideas endlessly without understanding or ever checking them. The dogmatist is enveloped by fog and disappears from the sight of history. I suspect that the vast majority of 'experts' are dogmatists. To go into mathematics is hard work, without pay or promotion. To admit to having no knowledge of mathematics is much more honest. It is up to the physicists to explain all the obscurity in order that enlightenment may prevail, otherwise there is no purpose to what they are doing. They could get bigger salaries in other jobs and the vast majority has no sufficient knowledge of mathematics. Some have no knowledge of knowledge itself. The word 'covariant' is horrible jargon. It means that the format of the covariant derivative is the same when that covariant derivative is transformed into another frame of reference. So what? One of my favourite enemies always used to fire off this question (or very common assault) without doing any work himself. The answer to those two little words is that covariance became a law of physics, a law of general relativity. It is one of the few things that have survived the scholarly criticism of the past decade, by my colleagues at AIAS and myself. Mr So What would fire off another round: physics? So what? He is long vanished into total obscurity, the archetypical career man.

The idea of the by now familiar vector appears to have been introduced in the latter part of the nineteenth century by a predecessor on the Civil List in Britain, Oliver Heaviside, who used vectors to simplify the earlier Maxwell equations of electrodynamics. Maxwell himself did not use vectors - he used Hamilton's quaternions, Hamilton being another predecessor on the Civil List. Gibbs also played a leading role in the development of vectors. I always refer to the equations of electrodynamics as the Maxwell Heaviside equations abbreviated to MH. These can be written as four vector equations, two are homogeneous, two are inhomogeneous. The homogeneous equations deal with the electromagnetic field itself, and are the Gauss law of magnetism and the Faraday law of induction. The two inhomogeneous equations deal with the interaction of the field with material matter, notably one electron. They are the Coulomb law and the Ampere

Maxwell law. The latter is named after Ampere and Maxwell because the latter modified the Ampere law with the Maxwell displacement current. These four laws were firmly based on experiments, and they first appeared as four laws in Heaviside's analysis. In Maxwell's earlier analysis there are many equations, much more difficult to apply and to understand than those of Heaviside. In addition to the four laws there are constitutive equations which introduce polarization and magnetization. For the purposes of this long essay, the key idea is that the MH equations are equations of special relativity. The modern era ideas of relativity originate in the MH equations. They are not equations of general relativity in the fogma (or 'received opinion', to be as polite as possible), but in ECE theory they are correctly relativistic in general, a well recognized triumph of ECE theory.

In the absence of a relatively simple and completely successful unified field theory such as the twenty first century Einstein Cartan Evans (ECE) unified field theory, the twentieth century wandered mistily along on two different legs, one for electrodynamics, the other for gravitation, the former being special and the latter general relativity. So physics prior to ECE became incomprehensible and meaningless, it was two different philosophies riddled with errors. It is miraculous how expensive experiments were designed to test these errors and come up with triumphant precision. This failing of human nature existed in ancient and mediaeval times, and has not changed one iota. It is a supplication to dense fog, one foot looking for the other. The idea of relativity emerged from the structure of the MH equations because Newtonian ideas do not apply in these equations. In Newtonian dynamics a velocity is added to another to give the sum of velocities. 'Common sense' in the seventeenth century. Common sense has a habit of being anthropomorphic, the sense is common because it is human, and by 'sense' what is really meant is fogma. In Newton's time, adding one linear velocity to another could go on indefinitely, so that the result could ultimately reach infinity. Something could go infinitely fast. A signal from earth could reach a distant star immediately, and in the terribly obscure, seventeenth century language of physics there could be action at a distance. Having based his work on this idea, Newton proceeded to reject it in that letter to Bentley: 'Pray do not ascribe etc.' Newton could be as dogmatic as anyone, and was not a particularly pleasant politician in his spare time from minting coins. In electrodynamics however it was realized by Maxwell that all that does not work, light has a speed c which is thought still to be a constant in the vacuum to an excellent approximation. The latest claims, made at incredible expense, are that it may vary by a tiny amount from c in the vacuum. Maxwell realized that light is made up of electromagnetic waves which travel at c in a vacuum, but not in a material. That was Maxwell's greatest contribution to physics in my own opinion. His equations are essentially unworkable, and it took a genius like Heaviside to unravel them at no expense at all. Maxwell realized that light does not travel faster than c. The first relativists, Heaviside, Fitzgerald and Lorentz, realized that it cannot travel faster than c, and threw Newton away.

The word ‘relativity’ came into fashion because of the Lorentz transform, needed to explain the results of the Michelson–Morley experiment. Although challenged by a few contemporary thinkers, and they may be right, the Michelson–Morley experiment is to most physicists reproducible and repeatable. In the Michelson–Morley experiment light is proven to be the same whether or not it travels with the earth’s rotation. The linear velocity v of the earth’s surface does not add to c, there is no v plus c. To the best of my knowledge, relativity as an idea started in correspondence between Fitzgerald and Heaviside, and after that with Lorentz. Neither Heaviside nor Fitzgerald had a clear way of putting their ideas into mathematics, but Fitzgerald wrote a half page article introducing the idea that was used almost immediately by Lorentz in his mathematical work. Fitzgerald proposed it in words, Lorentz put it into mathematical format. The most famous part of that correspondence is the Lorentz transform. In my opinion relativity began with the Lorentz transform although Heaviside came very close to the right answer. Fitzgerald got the right answer in words.

The transform deals with the relative properties of two frames of reference, and hence the words ‘relativity’ and ‘relativistic’. Considering two frames of reference can be terminally boring, but not to Lorentz. He must have played around a lot with different ideas before he hit on the right one, in real life as opposed to the textbooks this occurs very often almost by accident. Foremost in his mind was the constancy of c, in any transformation from one frame to another c must be constant because of the Michelson–Morley experiment. He was considering two frames of reference because Michelson and Morley had devised an experiment in which light was observed perpendicular and parallel to the earth’s direction of rotation. There could be a frame attached to the earth, our own human vantage point, but there could also be a frame outside the earth, a frame in which the earth could move. For example the earth moves with respect to an observer at the centre of the sun, which itself moves with respect to a frame at the centre of the Milky Way galaxy. Even so it took a tremendous leap of imagination for Lorentz to go any further. He had to dispense with common sense. He had to change the meaning of a frame of reference and he had to change the meaning of space itself. It could no longer be Newton’s three dimensional space unchanged for two thousand years from Euclid’s time. The space of Newton and Euclid is independent of time. The idea that Lorentz hit upon is that time, as well as distance, can vary from frame to frame, but c cannot. I can imagine a hypothetical correspondence with Newton’s wig falling off and ‘pray do not ascribe any more’ being hastily scribed to Lorentz.

In the comfortable, two thousand year old dogma (sorry, ‘received opinion’) the square of a radius vector is constant if the vector is rotated about the origin of the frame. To make this idea clearer, consider the case of rotation in the plane labelled by axes X and Y. Before the rotation, the vector is defined by components X and Y and basis or unit vectors, usually the familiar Cartesian unit vectors. After rotation, X and Y change, but the square of X added to the square of Y does not change because the length of the vector is not changed by a rotation. This familiar idea is the point from which Lorentz took off. He added the third dimension Z, and denoted the sum of X squared plus Y squared plus Z squared by *r* squared. Where does a constant c come in to the theory? The Lorentz transformation postulates that *c* squared *t* squared – *r* squared is constant in any transformation from one frame to another. This difference in terms, *c* squared *t* squared – *r* squared, is denoted *c* squared tau squared, and tau is named 'the proper time' in the admittedly terrible jargon. Therefore, time multiplied by *c* becomes a fourth coordinate, and space is transformed into spacetime. The position vector becomes four dimensional and has four components. It is denoted by (*ct*, X, Y, Z) instead of (X, Y, Z). This is known as a contravariant position vector. Minkowski later realized that an elegant formulation of special relativity can be achieved with the use of the covariant position vector (*ct*, –X, –Y, –Z). The key idea of special relativity is that time in one frame of reference is different from time in another. Special relativity reduces to one idea, that time can change from one frame to another, but *c* cannot. So there must be experiments that can be devised to test this idea.

The proper time tau is the square root of (1 – *v* squared / *c* squared) multiplied by *t*. If a particle is observed to be moving at a velocity v with respect to an observer, the latter measures a time interval delta *t* in the observer or laboratory frame *K*. The proper time interval measures the time in the frame at which the particle is at rest. For example the proper time interval is the time for a passenger on board a plane, the passenger (and his watch) is at rest with respect to the plane. The proper time interval is smaller than the observer time interval. So a watch on board a plane may record that one hour has passed, but for a watch on the ground, two hours have passed. Time on board the plane is slower than time on the ground, in other words the watch on board the plane runs slow. This phenomenon has been verified experimentally to astonishingly high precision, and this is meaningful relativity, worked out not by Einstein, but by Lorentz and others such as Voigt long before Einstein's work of 1905. The jargon is horrendous, and the phenomenon is known as 'time dilatation'. All it means is that the watch runs slow on board a plane. This type of relativity was developed elegantly by Lorentz and Poincare in the context of electrodynamics using the then new tensor methods of Levi-Civita, Ricci, Bianchi and others around 1900. A tensor is a kind of matrix, it can exist in one dimension as a row or column matrix and then becomes a vector. A matrix can exist in any number of dimensions.

One of the most elegant parts of the least fogmatic type of physics is the expression of the MH equations in tensor format. There is one homogeneous tensor equation and one inhomogeneous tensor equation. The electromagnetic field is expressed as a four by four antisymmetric matrix and is known as the electromagnetic field tensor. The elegance is revealed most clearly when one realizes that the Lorentz transform is worked in to the field tensor, which transforms into another frame using the transform twice (i.e. the transform applied to a tensor), and gives the Lorentz force equation. This formulation is therefore covariant, and a valid theory of special relativity. The field tensor is expressed in terms of the derivatives of the four potential, which combines the ideas of scalar and vector potential used by Heaviside. In this early theory, the Poincare Lemma can be used to show that there cannot be a magnetic monopole. Many scientists argue that a magnetic monopole can exist, but the experimental evidence is still debated. In ECE theory the existence of a magnetic monopole is left as an open question. All these great achievements were made when Einstein was still a student, a very bright one, confident with a clear grasp of concept, but prone to mathematical errors and impatience.

These early achievements of the first decade or so of special relativity (about 1888 to about 1900) were clear and precise, and were not confused by failed concepts such as gauge theory introduced by Weyl much later, and then elaborated into a completely failed particle theory. To Heaviside, the electromagnetic potential was real, and also to Faraday, another predecessor on the Civil List. The achievements of 1888 to 1900 did not involve Einstein at all. In his writings, Einstein confuses matters quite frequently, as anyone who has tried to read his original papers finds out. He also had a tendency to use work without citing it. For example the famous rest energy was worked out before Einstein and published in Italian, a language in which Einstein was fluent. It is very probable that he read this paper. In my opinion the lasting contributions by Einstein in special relativity rest on his development of the idea of relativistic momentum and relativistic kinetic energy, and his development of the momentum four vector. The two Einstein principles of the fogma should really be attributed to Lorentz and before him back to Maxwell.

In about 1900, Max Planck made his own leap of imagination in introducing the idea of the quantum of energy. Planck did not name this quantity the photon, the name arrived on the scene much later, as did the name 'quantum mechanics'. The physics before Planck had run into difficulties, notably the inability of the Rayleigh Jeans law to describe radiation over all frequencies from heated material, known in the densest of jargon as 'black body radiation'. There were also difficulties in explaining the photoelectric effect using the classical theory of Maxwell and Heaviside. The MH theory had been made into an elegant expression of special relativity, but this still did not explain the phenomena of what we know now as quantum physics. The dichotomy between relativistic and quantum physics remained a problem throughout the twentieth century and produced some of the densest fog in history, the Copenhagen interpretation. The latter has been revealed as hocus pocus many times, but as usual it is more comfortable to teach it rather than learn it. It produces such amazing nonsense that it tends to make fools out of all teachers. I remember this as a student myself, my teachers at Aberystwyth openly made fun out of it, then asked us students to regurgitate it in the examinations. This kept them in a job, and human nature again. Einstein contributed to the early quantum theory, known as 'the old quantum theory' by using the idea of quantum of energy (later named the 'photon') to explain the photoelectric effect. In the dogma this explanation is attributed wholly to Einstein, but with the advent of google it is found in about half an hour that others worked on it and that things are not at all clear cut. Nonetheless this is the work for which Einstein was awarded a Nobel Prize. He also devised the theory for absorption and emission and in 1906 tried to merge the ideas of special relativity and the old quantum theory by asserting that the photon (as we now know it) has mass.

This seems an obvious notion – as usual in retrospect – but more than a hundred years later the mass is still not known. In fact the AIAS work of 2010 and 2011 has revealed an astonishing lack of internal consistency in the most elementary kind of particle theory throwing the idea of fixed, unchanging, elementary particle mass into doubt. At incredible expense again, it has been found that the standard physics' particle theory has collapsed, it was all built around the mythical Higgs boson, and built in a very shaky way. I have been pointing this out for years. The electron had been known since Thomson's work, or was it Tesla's work? Soon there would be evidence for alpha particles from experiments by Rutherford, first at Manchester than at Cambridge. Einstein did again contribute with merit in 1905 by using the Brownian motion to prove the existence of molecules. The Brownian motion was discovered experimentally by another Civil List predecessor, Robert Brown, a Scottish botanist who discovered many thousands of new species of flora in Western Australia. He discovered the motion in pollen particles, which are macroscopic particles enormously larger and heavier than a molecule. The motion of the pollen particles seems to be random, not governed by Newton's ordered world of cause and effect. Many physicists a hundred years ago did not accept the atomic theory of another Civil List predecessor John Dalton, now about two hundred years old. Dalton would be described in our terms as a physical chemist. To chemists an atomic theory was proven by the periodic table of the elements, chemical reactions, valency, atomic number, and much else, but to physicists this was all dark matter or phlogiston, or horse hair. Einstein and others such as Langevin and Smoluchowski explained Brownian motion using a random or stochastic term in the Newton force law (the Langevin equation) or by using diffusion equations with stochastic terms and the ideas of Maxwell and Boltzmann in statistical mechanics. Einstein in 1905 used a diffusion equation and the idea borrowed from chemistry of the Avogadro number. It began to dawn on physicists that atoms and molecules really existed, a hundred years after Dalton.

The decade 1905 to 1915 is usually seen as the transition between special and general relativity. A principle of relativity is that the speed of light is constant in a vacuum, and is attributed by fogma solely to Einstein in 1905. However its origins go back to Maxwell as I have argued here already. The other principle of 1905 attributed to Einstein is that uniform motion is unobservable, which means that only the relative motion of inertial frames is observable and the concept of absolute rest is meaningless. The laws of physics are the same in all inertial reference frames. However, this is merely a restatement of the Lorentz transform. There are experimental claims that neither principle holds, there are claims that c is not constant and that the Michelson Morley experiment does not produce a null result. The fogma claims that in 1906, the idea of mass being equivalent to energy was proposed by Einstein, and that that led to the rest energy equation $E = mc^2$. Contemporary scholarship has shown that this famous equation was proposed before Einstein in an Italian journal, as discussed already. In 1907 however Einstein made two genuine contributions, in the theory of specific heats and in the transverse Doppler effect. His specific heat theory showed that the law E is equal to h nu is a general law, and not one restricted to black body radiation. The concept of photon as particle is also due to Einstein, and not Planck, the latter's idea being restricted to the quantum of electromagnetic energy. The fogma claims that 1907 saw the first derivation of $E = mc^2$, but this is now known to be false. It was derived by another scientist near the turn of the century as argued already. The fogma claims that in 1907 Einstein proposed the equivalence principle, derived light bending due to gravitation and the gravitational red shift. How could he have done that without a field equation and with the ideas of general relativity not yet crystallized even in his own mind? The ideas could only have been heuristic in the year 1907.

In 1911 Einstein proposed that special relativity and the equivalence principle hold locally, but in 1912 realized that the Lorentz transform must be generalized. This idea of 1912 is his major contribution in my opinion, because it introduces a geometrical theory of relativity – general relativity. In contemporary language the Lorentz transform becomes a special case of the general coordinate transform. The idea of gravitation being a field tensor was introduced with Marcel Grossmann in 1913 but no field equation was given in 1913. In 1914 Einstein introduced a theory based on the geodesic motion of point particles. He had been fully aware of Riemann geometry for some time, and regularly corresponded with Levi-Civita, who had introduced the idea of curvature. It becomes clear, however, that the key concept of torsion was never used in the decade 1905 to 1915 – indeed it had not been thought of by any mathematician. It was Cartan who first pointed out in the early twenties that geometry must be characterized both by curvature and torsion, in the two Cartan structure equations. In four papers of late 1915, the fatal errors were made. The Einstein field equation appeared in November in pages 844 to 847 of that year's proceedings of the Royal Prussian Academy. A paper on the precession of Mercury appeared in pages 831–839, also in November 1915, and was shown to be incorrect by Schwarzschild in December 1915. Recent scholarship (available on the web) has translated Schwarzschild's criticism in his letter to Einstein of December 1915, leaving no doubt as to the incorrectness of the calculation.

The paper on the field equation, pages 844 to 847, used one idea from geometry and one from physics. The idea from geometry is the second identity of Bianchi. I have shown during the course of developing ECE theory that the second Bianchi identity rests on the incorrect assumption that there is no spacetime torsion. It was entirely natural for Bianchi to have worked without torsion, because it was not known to him. It took Cartan's genius to realize its existence, almost twenty years after Bianchi. The error made by Einstein was therefore the incorrect omission of torsion. The rest of twentieth century general relativity compounded this error, and is meaningless relativity. In 2003 the ECE theory started to include torsion and this led to a straightforward unification of physics. The idea from physics used by Einstein was the Noether Theorem, which is based on the canonical energy momentum tensor. Einstein made the second Bianchi identity proportional to the Noether Theorem through a constant k. Both the Bianchi identity and the Noether Theorem use the covariant derivative, but Einstein assumed that the quantities being differentiated covariantly could be made proportional. There appears to be no way of proving this assumption, which leads to the incorrect Einstein field equation of general relativity in which a geometrical tensor of Bianchi is made proportional to the canonical energy momentum tensor, both being symmetric tensors. The tensor of Bianchi is known as the Einstein field tensor, the gravitational field of force is thought of as geometry.

Schwarzschild must have known of the field equation prior to its publication in November 1915, because he heavily criticised Einstein in a letter of December 1915 in which he declared 'friendly war', and in which he gave his solution to the field equation as mentioned already in this long essay. The rest of the fogma is entirely meaningless because it too neglects torsion. For example in 1917 Einstein is supposed to have initiated a cosmology based on big bang, one of the densest occurrences of low visibility ever to threaten enlightenment. The truth is that Einstein himself rejected big bang in 1939 as being unphysical. In 1918 Einstein is supposed to have introduced gravitational waves which are supposed to have been observed 'indirectly' in binary pulsars. The truth is that gravitational waves from a mathematically incorrect equation cannot exist in nature. Einstein's search for a unified field theory was flawed from the very beginning because of his neglect of torsion. He must surely have known of the existence of torsion from his correspondence with Cartan in the twenties. However Einstein was already famous and trapped in his own fogma. Whatever the reason, he did not use Cartan geometry as given by Cartan. The latter provided the geometrical basis for a unified field theory in the early twenties – all that is needed for a unified field theory. In 1929 Einstein used Hamilton's principle in an another failed attempt at unification, and in 1950, towards the end of his life, considered the first and second Bianchi identities in a generalized theory of gravitation, but again neglected torsion. The true identity with torsion was given by Cartan and is used in ECE theory.

The search for unification by Einstein is a long and heroic one, but was doomed from the outset for the reasons just given, and also because he tried to merge the Maxwell Heaviside (MH) equations with his own, sadly incorrect, field equation. This attempt resulted in the Maxwell Einstein equations as they are known, but again these neglect torsion and are meaningless. In ECE theory the MH equations are written in a spacetime with torsion and curvature, and themselves become equations of general relativity. In ECE theory dynamics, gravitation and electrodynamics are based on Cartan geometry, and are based on torsion. In retrospect it is obvious that Einstein neglected the key and central concept of torsion. Once reinstated the unification of physics becomes relatively easy with a little bit of imagination. It is also easy to see why the initial fogmatic reactions to ECE were as they were – unification became maddeningly obvious to those tied up in strings and allergic to Ockham's Razor. Although Einstein himself had suggested photon mass in about 1905 or 1906, his twentieth century successors became bogged and befogged by the idea of the photon as a particle having no mass. This led to all kinds of difficulties that were so severe that any sensible scholar would have abandoned the theory, and would have gotten by without the funding for the sake of dignity and self respect. One of these ghastly fogmatic failures was the assertion that electromagnetic radiation in the vacuum could only have transverse polarizations, defying the existence of the third dimension for the sake of gauge invariance. Mass was supposed to have been given to elementary particles by a magical process based on what was known incorrectly as the Higgs boson. In truth it was devised by others as well as Higgs, but was in any case a failure. This year CERN has at last admitted that it does not exist, leaving us all a lot poorer, not intellectually, but in terms of extracted taxation.

In Schwarzschild's letter of December 1915 to Einstein, the former proposed a line element solution to the field equation proposed in that year by Einstein. This solution is not that commonly called the Schwarzschild metric, and in the same letter, Schwarzschild heavily criticised the perihelion calculation by Einstein. Unfortunately Schwarzschild died only a few months later, in 1916. Einstein was never able to refute the criticism by Schwarzschild, and published only one short note thereafter on perihelion precession. He never returned to the subject after that. This does not give an objective scholar much confidence in the subject of general relativity without torsion. Historical scholarship alone is enough therefore to show that neither Einstein nor Schwarzschild correctly predicted the precession of the perihelion of Mercury. The great disservice done to science resides in the endless repetition of the contrary claim that they did. Schwarzschild did not infer the line element solution commonly attributed to him. All that has to be done to confirm this is to read the two papers actually published by Schwarzschild on the subject, in 1916, and to read the English translation on the web of his December 1915 letter to Einstein.

The impression given of the years 1915 onwards to the early twenties is also negative in a historical context. It is not clear why the original solution by Schwarzschild of December 1915 was changed, and why the changed solution was attributed falsely to him. The changed solution contains a singularity at the origin and is not a well behaved mathematical function for this reason. It seems to have been chosen in order to force the Einstein theory to reduce to the Newton theory. That procedure is not objective science, which compares theory with experimental data, not with another theory. The experiment by Eddington and colleagues carried out on May 29th 1919 was based on two papers by Einstein. One of these was published in Annalen der Physik 35, 898–908 (1911), and the other in Proceedings of the Royal Prussian Academy of 1915(2), pages 831–839. This is also the paper in which the incorrect perihelion precession calculation was given. So Schwarzschild's criticisms hold also for the light deflection calculation of Einstein, which was also incorrect. In UFT150 of the unified field theory series on www.aias.us the incorrectness of the light deflection calculation was shown in several ways. In UFT194 the basis for the Einstein theory was refuted conclusively in a simple way, using algebra checked by computer. In the 1911 paper Einstein produced a result for light deflection which was half that produced in the 1915 paper. Neither of these papers was refereed. In 1919 Eddington did not have the precision to test the claims by Einstein, and unfortunately a given data set was chosen which happened to coincide with the 1915 prediction. It is now possible to measure the light deflection accurately, but it cannot be due to Einstein's incorrect theory.

Shortly after the Eddington experiment had subjectively favoured Einstein by choice of data set, Cartan and colleagues inferred spacetime torsion and showed that the geometry used by Einstein is incorrect because of its incorrect omission of torsion. Einstein was made aware of this by Cartan as discussed already, but the false impression given by the Eddington experiment meant that his curvature theory was adopted uncritically. For many years general relativity remained a relatively obscure part of physics until it was somewhat artificially elevated into cosmology by the big bang theory. The term 'big bang' was coined by Fred Hoyle as a joke. He rejected the theory by Hawking and Penrose and others, and Einstein himself had rejected it long before Hoyle, in 1939. It is a typical media hype of the twentieth century brought about by a desire for fame and notoriety, not a desire for science and scholarship. By the time I became an undergraduate in 1968 I had heard of the big bang theory vaguely, but gave it little credence. I had not heard of black hole theory at all. This was another lurid term, coined by the otherwise eloquent John Wheeler, Einstein's assistant. My rigorous examination of the claims of Einsteinian general relativity (EGR) began in earnest in 2003, the year in which ECE theory was inferred. I noticed the existence of spacetime torsion in chapter three of Sean Carroll's book: 'Spacetime and Geometry: an Introduction to General Relativity' (Addison Wesley, New York, 2004 and online notes of 1997). The ECE theory is based on torsion, and is the first generally covariant unified field theory built in this way. So from the outset I decided that torsion cannot be thrown out with the bathwater. Reading around, it soon became apparent that the EGR was built on the axiomatic or arbitrary suppression of this basic property of geometry. Torsion had been neglected in the early years of geometry because it was not known. Christoffel had arbitrarily defined the connection as being symmetric in about 1869, and it took until about 1922 to realize that it is not symmetric in general. A symmetric connection means zero torsion. Unfortunately for EGR and science, Einstein adopted the symmetric connection in the years 1905 to 1915 for the same reason, it was the only connection of geometry known to mathematics.

A lot of my work in the early development of ECE theory consisted of giving proofs with details left out by Carroll, as exercises to numbed graduate students at Harvard and elsewhere. These were proofs of the tetrad postulate, the two Cartan structure equations and the Cartan identity. Later I inferred my own identity using Hodge duals in four dimensions and named it the Evans identity to distinguish it from the Cartan identity. In fact the Evans identity is an example of the Cartan identity. All these proofs check themselves as being correct. So I did not leave anything as the proverbial 'exercise for the student'. In order to learn Cartan geometry thoroughly a mastery of all these proofs is needed, together with a mastery of differential forms, tensors and vectors. There can be no valid criticism of ECE theory because it is based on this well tried geometry and all these rigorous proofs, each fortifying the other. One of the most important of them is the derivation of the structure of torsion and curvature using the commutator of covariant derivatives. This method shows that curvature is always accompanied by torsion, which is why there are two Cartan structure equations, one for the torsion form, the other for the curvature form. Up until paper 122 of the unified field theory series I still accepted the old dogma that the connection is in general asymmetric, or without any particular symmetry. It is amusing to note at this stage of enlightenment that this dogma contradicted the other dogma about the Christoffel connection being symmetric. In fact neither dogmatic assertion is true, the connection is now known to be antisymmetric, a finding that is alone enough to refute EGR, with its symmetric connection, a throw back to Christoffel in 1869.

The proof of connection antisymmetry is simple, it is a consequence of the antisymmetry of the commutator. Subsequently I have given many fortifying proofs of the first one in UFT 122 on www.aias.us. Earlier I had given the correct version of the old second Bianchi identity in UFT 88, now an intensely studied paper. The Einstein field equation was based directly on the old (torsionless) second Bianchi identity. It gradually became very clear to me that the EGR is not only wrong, but wildly wrong. So I began to look for other manifestations of this ghastly fate of twentieth century cosmology. Naturally I did not make myself popular, (except with Queen Elizabeth), but so what, as Mr So What would have fired out. I gradually hit upon the idea of using the conical sections, but modified with a precession factor that I have named x, for the sake of blandness and old times at school, when x saturated all algebra. The innocent factor x was to have spectacular consequences that finished off EGR there and then and now.

It works simply by multiplying the angle in the polar representation of the conical sections. This might be thought to be terminally boring, but as so often in real discovery, it contains profundities of hidden truth, hidden for almost two thousand five hundred years since the Greeks discovered the conical sections: the circle, ellipse, parabola and hyperbola. The terrible failure of EGR became apparent simply by differentiating the conical section with x for an ellipse. This was first done in UFT 202. At that point I called this type of ellipse the precessing ellipse, because x very close to unity makes the ellipse turn on its focal point, or precess as the Latin has it. The equivalent result from EGR is indeed wildly wrong, and there is no denying it. A number of previous essays and broadcasts in this series have dealt with this result in hopefully understandable terms, and judging by their popularity in English and Spanish on www.aias.us, they have succeeded very well. Robert Cheshire narrates them to a high professional standard, as does Alex Hill in Spanish translation.

I also noticed a few months ago that the precessing ellipse gives a different force law both to that of Newton and of Einstein. This turned out to be the truly universal law of gravitation and can be obtained with the wonderful elegance of Lagrangian dynamics (Lagrange 1788). EGR gives the wrong force law for a precessing ellipse, and this terrible blunder confirms the foreboding by Karl Schwarzschild in his letter already mentioned in this long essay, dated Dec. 22nd., 1915. No wonder Einstein never returned to the subject. For nearly a century cosmologists have laboured mightily to prove incorrect mathematics. My co author Horst Eckardt used the computer to finish off EGR by showing that the incorrect EGR force law produces a wild mess of an orbit, not at all a precessing ellipse. Within a few months I derived many other ways of refuting EGR but at the same time started to redevelop metric based general relativity from the beginning.

The latest discovery made by AIAS was made in a ridiculously simple way by allowing x to vary over its full range for every type of conical section, whereupon appear a vast array of new conical sections of many types, all of them having fractal properties. This discovery was totally unexpected and occurred when x was varied over a small range to show the effect on a hyperbolic orbit. The intent being to demonstrate non Newtonian light bending without use of EGR. Out of curiosity I asked Horst Eckardt to vary x over its complete range, and the hyperbola took on a life of its own as explained in UFT 215. Very quickly, Horst Eckardt and Ray Delaforce produced the wonderful fractal conical sections displayed on the blog of www.aias.us. For several hundred years the conical sections have been synonymous with orbits, so we can expect many new orbits to be discovered. All are produced by the new universal gravitational potential. This has the same format exactly as the potential of the Schroedinger equation in atoms and molecules, so as Gareth Evans pointed out, is universal on all scales. The EGR is wholly unable to produce this new subject in mathematics and physics.

This kind of discovery has an immediacy that makes it clear to everyone, whatever their prior ideas. The most vivid analogy is the discovery of a hoard of shining and brilliantly crafted gold in archaeology, one that lain buried for over two thousand five hundred years. I can do no better than to let the new patterns impart their life and beauty to the beholder's vision. We have posted the very first patterns to emerge on the blog of www.aias.us. To be absolutely fair, a sixth former called David Herbert at Cantonian School in Cardiff wrote me a letter earlier this year, sketching out some of the petal orbits. He was puzzled that they did not look like precessing ellipses. So he too must be given his fair share of credit. He had accidentally used a value of x that was too large for a precessing ellipse, for which x must be very close to unity. So thanks to the teachers and head teacher at Cantonian school for their help and enlightenment.

Myron W Evans

Craig Cefn Parc, April 2012

A cosmology based exclusively on torsion

M.W. Evans[1*], H. Eckardt[2] and D.W. Lindstrom[3*]**

**Alpha Institute for Advanced Studies (AIAS) (www.aias.us); **Unified Physics Institute of Technology (UPITEC) (www.upitec.org)*

The metric compatibility condition of Riemann geometry is used as the basis of a new method of finding the antisymmetric connection from a given metric. This method shows that the force of gravitation is directly proportional to Riemann torsion, and that all elements of the Riemann curvature vanish. Computer algebra is used to produce examples of antisymmetric connection elements. The Einstein field equation is nowhere used in this new cosmology, based on a rigorously correct geometry with considerations of metric compatibility and the correct Bianchi identity. All cosmology is reduced to spacetime torsion, and spacetime curvature eliminated from the subject.

Keywords: ECE theory, torsion based cosmology.

1. Introduction

In recent publications [1–10] the use of a symmetric connection in cosmology has been shown to be incorrect geometrically. The commutator method [11], well known in geometry, shows that the symmetric connection implies a symmetric commutator, so that the Riemann curvature and torsion both vanish. The only physically meaningful connection must be antisymmetric in its lower two indices. This means that the spacetime torsion is always non-zero and that the Einstein field equation is incorrect [1–10] irretrievably. In Section 2 it is shown that the antisymmetric connection can be obtained straightforwardly using the equation of metric compatibility. For a diagonal metric the connections are obtained very simply by hand calculation. For non diagonal metrics, computer algebra is used in Section 3 to deduce the connections. Only one equation of metric compatibility is needed. In the early days of tensor analysis three such equations were used in cyclic permutation to give a well known [11] expression for the

[1]e-mail: EMyrone@aol.com
[2]e-mail: horsteck@aol.com
[3]e-mail: dwlindstrom@gmail.com

symmetric connection in terms of metric derivatives, but no account was taken of the commutator of covariant derivatives, which shows that the connection must be antisymmetric. An important result of this paper is that all elements of the Riemann curvature vanish, so that cosmology becomes a subject based exclusively on torsion. This is a major feature of the post Einstein paradigm shift. The antisymmetric connection and the force of universal gravitation are shown to be directly proportional through the rest energy of the attracted particle.

2. The connections from a diagonal metric

The metric compatibility condition is [1–10]:

$$D_\rho g_{\mu\nu} = \partial_\rho g_{\mu\nu} - \Gamma^\lambda_{\rho\mu} g_{\lambda\nu} - \Gamma^\lambda_{\rho\nu} g_{\mu\lambda} = 0 \tag{1}$$

where $g_{\mu\nu}$ is the symmetric metric and $\Gamma^\lambda_{\rho\mu}$ the Christoffel connections. Consider a metric in cylindrical polar coordinates in a plane of the type:

$$g_{\mu\nu} = \begin{bmatrix} 1-\frac{r_0}{r} & 0 & 0 & 0 \\ 0 & -\left(1-\frac{r_0}{r}\right)^{-1} & 0 & 0 \\ 0 & 0 & -r^2 & 0 \\ 0 & 0 & 0 & -r^2\sin^2\phi \end{bmatrix} \tag{2}$$

where:

$$r_0 = \frac{2MG}{c^2}. \tag{3}$$

Here M is the mass of the attracting object, c the vacuum speed of light and G the Newton constant. This metric is one possible solution of the ECE Orbital Theorem of UFT 111 (www.aias.us). It gives an accurate description of the relativistic Kepler problem [12].

For

$$\mu = \nu = 0 \tag{4}$$

Equation (1) becomes:

$$\partial_1 g_{00} - \Gamma^0_{10} g_{00} - \Gamma^0_{10} g_{00} = 0 \tag{5}$$

so

$$\Gamma^0_{10} = -\Gamma^0_{01} = \frac{r_0}{2r^2\left(1 - \frac{r_0}{r}\right)}. \tag{6}$$

It is already clear that the connection can be obtained from the metric very easily. Yet this fact has been overlooked for 110 years. Similarly the non vanishing connections are:

$$\Gamma^1_{11} = \frac{1}{2g_{11}} \partial_1 g_{11}, \tag{7}$$

$$\Gamma^2_{12} = \Gamma^3_{13} = \frac{1}{r}, \quad \Gamma^3_{23} = \cot\phi. \tag{8}$$

It is seen that there is one symmetric connection Γ^1_{11}, but as follows this connection results in zero torsion and curvature, so plays no part in universal gravitation. Consider the well known commutator equation of geometry [11]:

$$\left[D_\mu, D_\nu\right] V^\rho = -\left(\Gamma^\lambda_{\mu\nu} - \Gamma^\lambda_{\nu\mu}\right) D_\lambda V^\rho + R^\rho_{\sigma\mu\nu} \tag{9}$$

where D_μ is the covariant derivative, V^ρ vector of any spacetime of any dimension, and where the Riemann torsion and curvature are defined respectively by:

$$T^\lambda_{\mu\nu} = \Gamma^\lambda_{\mu\nu} - \Gamma^\lambda_{\nu\mu} \tag{10}$$

and

$$R^\rho_{\sigma\mu\nu} = \partial_\mu \Gamma^\rho_{\nu\sigma} - \partial_\nu \Gamma^\rho_{\mu\sigma} + \Gamma^\rho_{\mu\lambda}\Gamma^\lambda_{\nu\sigma} - \Gamma^\rho_{\nu\lambda}\Gamma^\lambda_{\mu\sigma}. \tag{11}$$

It is seen immediately from Eq. (9) that when:

$$\mu = \nu = 1 \tag{12}$$

the commutator, symmetric connection, torsion and curvature all vanish. It follows that the equation of metric compatibility (1) applies only to antisymmetric connections, i.e. the connection of any spacetime of any dimension is antisymmetric. Otherwise there is a contradiction between two fundamental equations (1) and (9).

From Eq. (6) it is seen that in the limit:

$$r \gg r_0 \tag{13}$$

the inverse square law of gravitation is given by:

$$F = \frac{-mMG}{r^2} = -mc^2 \Gamma^0_{10} \tag{14}$$

so universal gravitation is due exclusively to torsion. This is a central feature of the post Einstein paradigm shift. Table 1 summarises the results of this section. All elements of curvature vanish. Curvature plays no role in universal gravitation, a second major feature of the post Einstein paradigm shift.

Table 1: Non - Zero Connection and Torsion Elements of Metric (2).

Antisymmetric Connection	Torsion
$\Gamma^0_{10} = -\Gamma^0_{01} = \dfrac{r_0}{2r(r-r_0)}$	$T^0_{10} = 2\Gamma^0_{10}$
$\Gamma^2_{12} = -\Gamma^2_{21} = \dfrac{1}{r}$	$T^2_{12} = -T^2_{21} = 2\Gamma^2_{12}$
$\Gamma^3_{13} = -\Gamma^3_{31} = \dfrac{1}{r}$	$T^3_{13} = \partial\Gamma^3_{13}$
$\Gamma^3_{23} = -\Gamma^3_{32} = \tan\phi$	$T^3_{23} = 2\Gamma^3_{23}$

3. Results for various diagonal metrics

In this section we discuss three examples of metrics and then specify the general solution method for diagonal metrics.

3.1. Examples for antisymmetric connections

Three additional metrics have beenanalyzed by computer algebra:

- Crothers metric with generalized Schwarzschild parameters,
- general spherical metric,
- spherically symmetric metric with perturbation.

These are derivable from the theorem of orbits of UFT paper 111 and therefore do not rest upon the obsolete Einstein equation. A detailed analysis with conventional (symmetric) Christoffel connections is given in [1]. In the following we give the results for the antisymmetric Christoffel connections. Two independent computer algebra systems were used to ensure correctness of the results. For the **Crothers metric with Schwarzschild parameters** these are:

$$g_{\mu\nu}=\begin{bmatrix}-\sqrt{\left(|r_0-r|^n+\alpha^n\right)^{\frac{2}{n}}A} & 0 & 0 & 0\\ 0 & \sqrt{\left(|r_0-r|^n+\alpha^n\right)^{\frac{2}{n}}B} & 0 & 0\\ 0 & 0 & \left(|r_0-r|^n+\alpha^n\right)^{\frac{2}{n}} & 0\\ 0 & 0 & 0 & \left(|r_0-r|^n+\alpha^n\right)^{\frac{2}{n}}\sin^2\vartheta\end{bmatrix} \tag{15}$$

$$\Gamma^{0n}_{10}=-\frac{|r_0-r|^n}{2(r_0-r)\left(|r_0-r|^n+\alpha^n\right)} \tag{16}$$

$$\Gamma^{1}_{11}=\frac{|r_0-r|^n}{2(r_0-r)\left(|r_0-r|^n+\alpha^n\right)} \tag{17}$$

$$\Gamma^{2}_{12}=-\frac{|r_0-r|^n}{(r_0-r)\left(|r_0-r|^n+\alpha^n\right)} \tag{18}$$

$$\Gamma^{3}_{13}=-\frac{|r_0-r|^n}{(r_0-r)\left(|r_0-r|^n+\alpha^n\right)} \tag{19}$$

$$\Gamma^{3}_{23}=\frac{\cos\vartheta}{\sin\vartheta} \tag{20}$$

The results for the **general spherical metric** are as follows. α and β are functions of r and t.

$$q_{\mu\nu}=\begin{bmatrix}-e^{2\alpha} & 0 & 0 & 0\\ 0 & e^{2\beta} & 0 & 0\\ 0 & 0 & r^2 & 0\\ 0 & 0 & 0 & r^2\sin^2\vartheta\end{bmatrix} \tag{21}$$

$$\Gamma^{0}_{00}=\frac{d}{dt}\alpha \tag{22}$$

$$\Gamma^0_{10} = \frac{d}{dr}\alpha \tag{23}$$

$$\Gamma^1_{01} = \frac{d}{dt}\beta \tag{24}$$

$$\Gamma^2_{11} = \frac{d}{dr}\beta \tag{25}$$

$$\Gamma^2_{12} = \frac{1}{r} \tag{26}$$

$$\Gamma^3_{13} = \frac{1}{r} \tag{27}$$

$$\Gamma^3_{23} = \frac{\cos\vartheta}{\sin\vartheta} \tag{28}$$

In some cases symmetric Christoffel symbols appear, these have to be discarded for torsion as described in the first section. In all cases the Riemann tensor is zero. This is an important difference to the results for the symmetric connections listed in [1]. Torsion alone is responsible for gravitation.

3.2. The general solution for a diagonal metric

The equation for metric compatibility (1) has three independent indices, this gives $4^3 = 64$ equations in combination. In case of a diagonal metric the number of index conbinations on the left hand side is reduced to 16, but at the right hand side metric elements other than $g_{\mu\mu}$ appear so that there are terms also for $\mu \neq \nu$. In the latter case linear equations arise. The coordinates are denoted by $x_0 \ldots x_3$ here. For example, the second five out of the 64 equations read:

$$Eq.(6): 2\,\Gamma^1_{01}\, g_{11} = \frac{\partial}{\partial x_0} g_{11} \tag{35}$$

$$Eq.(7): \Gamma^2_{01}\, g_{22} + \Gamma^1_{02}\, g_{11} = 0 \tag{36}$$

$$Eq.(8): \Gamma^3_{01}\, g_{33} + \Gamma^1_{03}\, g_{11} = 0 \tag{37}$$

$$Eq.(9): \Gamma^2_{00}\, g_{22} + \Gamma^0_{02}\, g_{00} = 0 \tag{38}$$

$$Eq.(10): \Gamma^2_{01}\, g_{22} + \Gamma^1_{02}\, g_{11} = 0 \tag{39}$$

As has been concluded from the examples for gravitation, the most important Christoffel connections follow from the case $\mu = \nu$. Then the following 16 equations appear:

$$Eq.(1):\ 2\,\Gamma^{0}_{00}\,g_{00} = \frac{\partial}{\partial x_0}\,g_{00} \tag{40}$$

$$Eq.(2):\ 2\,\Gamma^{1}_{01}\,g_{11} = \frac{\partial}{\partial x_0}\,g_{11} \tag{41}$$

$$Eq.(3):\ 2\,\Gamma^{2}_{02}\,g_{22} = \frac{\partial}{\partial x_0}\,g_{22} \tag{42}$$

$$Eq.(4):\ 2\,\Gamma^{3}_{03}\,g_{33} = \frac{\partial}{\partial x_0}\,g_{33} \tag{43}$$

$$Eq.(5):\ 2\,\Gamma^{0}_{10}\,g_{00} = \frac{\partial}{\partial x_1}\,g_{00} \tag{44}$$

$$Eq.(6):\ 2\,\Gamma^{1}_{11}\,g_{11} = \frac{\partial}{\partial x_1}\,g_{11} \tag{45}$$

$$Eq.(7):\ 2\,\Gamma^{2}_{12}\,g_{22} = \frac{\partial}{\partial x_1}\,g_{22} \tag{46}$$

$$Eq.(8):\ 2\,\Gamma^{3}_{13}\,g_{33} = \frac{\partial}{\partial x_1}\,g_{33} \tag{47}$$

$$Eq.(9):\ 2\,\Gamma^{0}_{20}\,g_{00} = \frac{\partial}{\partial x_2}\,g_{00} \tag{48}$$

$$Eq.(10):\ 2\,\Gamma^{1}_{21}\,g_{11} = \frac{\partial}{\partial x_2}\,g_{11} \tag{49}$$

$$Eq.(11):\ 2\,\Gamma^{2}_{22}\,g_{22} = \frac{\partial}{\partial x_2}\,g_{22} \tag{50}$$

$$Eq.(12):\ 2\,\Gamma^{3}_{23}\,g_{33} = \frac{\partial}{\partial x_2}\,g_{33} \tag{51}$$

$$Eq.(13):\ 2\,\Gamma^{0}_{30}\,g_{00} = \frac{\partial}{\partial x_3}\,g_{00} \tag{52}$$

$$Eq.(14):\ 2\,\Gamma^{1}_{31}\,g_{11} = \frac{\partial}{\partial x_3}\,g_{11} \tag{53}$$

$$Eq.(15):\ 2\,\Gamma^{2}_{32}\,g_{22} = \frac{\partial}{\partial x_3}\,g_{22} \tag{54}$$

$$Eq.(16):\ 2\,\Gamma^{3}_{33}\,g_{33} = \frac{\partial}{\partial x_3}\,g_{33} \tag{55}$$

3.3 The general solution for non-diagonal metrics

As soon as non-diagonal elements in the metric are present, the equation set becomes much more complex. Numerical methods for solution have to be applied then. The resulting equation set can be computed by computer algebra. We did this in the simplest case, adding two non-diagonal elements

$$g_{01} = g_{10}. \tag{56}$$

This already increases the complexity significantly. The five exemplary equations (35)–(39) take the form:

$$Eq.(6) : 2\Gamma^1_{01}\, g_{11} + \Gamma^0_{01}\, g_{10} + \Gamma^0_{01}\, g_{01} = \frac{\partial}{\partial x_0}\, g_{11} \tag{57}$$

$$Eq.(7) : \Gamma^2_{01}\, g_{22} + \Gamma^1_{02}\, g_{11} + \Gamma^0_{02}\, g_{10} = 0 \tag{58}$$

$$Eq.(8) : \Gamma^3_{01}\, g_{33} + \Gamma^1_{03}\, g_{11} + \Gamma^0_{03}\, g_{10} = 0 \tag{59}$$

$$Eq.(9) : \Gamma^2_{00}\, g_{22} + \Gamma^1_{02}\, g_{10} + \Gamma^0_{02}\, g_{00} = 0 \tag{60}$$

$$Eq.(10) : \Gamma^2_{01}\, g_{22} + \Gamma^1_{02}\, g_{11} + \Gamma^0_{02}\, g_{01} = 0 \tag{61}$$

The rest of the equation set looks similar.

References

1. M.W. Evans, S. Crothers, H. Eckardt and K. Pendergast, "Criticisms of the Einstein Field Equation". (Cambridge International Science Publishing, 2011).
2. M. W. Evans, H. Eckardt and D. W. Lindstrom, "Generally Covariant Unified Field Theory" (Abramis, 2005 onwards), in seven volumes.
3. Kerry Pendergast, "The Life of Myron Evans" (Cambridge International Science Publishing, 2011).
4. M.W. Evans, ed., Journal of Foundations of Physics and Chemistry (Cambridge International Science Publishing, June 2011 onwards).
5. The ECE sites: www.webarchive.org.uk, www.aias.us, www.atomicprecision.com, www.upitec.org, www.et3m.net.
6. L. Felker, "The Evans Equations of Unified Field Theory" (Abramis, 2007).
7. M.W. Evans, ed., "Modern Nonlinear Optics" (Wiley 2001, 2nd edition) in three volumes; ibid., M. W. Evans and S. Kielich (eds.), (Wiley, 1992, 1993, 1997, 1st edition), in three volumes.
8. M.W. Evans and L.B. Crowell, "Classical and Quantum Electrodynamics and the B(3) Field" (World Scientific, 2001).
9. M.W. Evans and J.-P. Vigier, "The Enigmatic Photon" (Kluwer, 1994 to 2002), in ten volumes hardback and softback.
10. M.W. Evans and A.A. Hasanein, "The Photomagneton in Quantum Field Theory" (World Scientific, 1994).
11. S.P. Carroll, "Spacetime and Geometry: an Introduction to General Relativity" (Addison Wesley, New York, 2004).
12. J.B. Marion and S.T. Thornton, "Classical Dynamics" (HB College Publishing, New York, 1988, 3rd. Edition).

Self consistent metric from torsional cosmology

M.W. Evans[1*], H. Eckardt[2] and D.W. Lindstrom[3*]**

**Alpha Institute for Advanced Studies (AIAS) (www.aias.us); **Unified Physics Institute of Technology (UPITEC) (www.upitec.org)*

It is argued that the basic theorem of Riemann geometry must be the commutator theorem that establishes the antisymmetry of the Christoffel connection. It is shown that for a diagonal metric all the elements of the Riemann tensor vanish, so that in the solar system for example the motions of the planets are governed by torsion alone. The antisymmetric connection is shown to be the only one that leads to self consistency between metric compatibility and the Evans identity. Using these rigorous criteria severely limits the number of self consistent metrics available for the development of cosmology.

Keywords: Torsional cosmology, self consistent metric, the fundamental commutator theorem.

1. Introduction

Recently in this series of papers [1–10] it has been shown that the Christoffel connection can be deduced straightforwardly by use of a single metric compatibility equation. The metric compatibility equation is one of the fundamental equations of Riemann geometry but the commutator equation is used to deduce the structure of the Riemann curvature and torsion. The correct use of the commutator equation [11] shows that the connection is always antisymmetric. Therefore the only valid equation of metric compatibility is one in which the connection is antisymmetric. By use of the Evans identity [12] it is shown in Section 2 that the use of an incorrectly non-zero symmetric connection leads to a self contradiction in Riemann geometry itself. This simple reasoning overturns 110 years of received opinion based on the symmetric connection and for a diagonal metric, makes cosmology into a subject based entirely on spacetime torsion. A self consistent metric is derived based on a combination of the metric compatibility condition and Evans identity in a spherically symmetric spacetime. The 'Schwarzschild' metric can

[1]e-mail: EMyrone@aol.com
[2]e-mail: horsteck@aol.com
[3]e-mail: dwlindstrom@gmail.com

be self consistent only in the limit of infinite separation between gravitating objects and was not in fact derived by Schwarzschild. It is unclear why it has been misattributed to him for nearly a century.

The concept of metric was introduced by Riemann and the concept of connection by Christoffel. In about 1900 Levi-Civita developed the idea of symmetric connection without use of the commutator equation, which shows that the connection is antisymmetric. So from the outset the received opinion of Riemann geometry was based incorrectly on the symmetric connection, or equivalently the neglect of torsion. In consequence of the erroneous use of a symmetric connection, three metric compatibility conditions were used to obtain what was asserted to be a uniquely defined symmetric connection. This was known as the fundamental theorem of Riemann geometry. The very fundamental commutator method on the other hand shows that the connection is always antisymmetric, so the fundamental theorem of Riemann geometry is incorrect and self inconsistent. The commutator method uses the commutator of covariant derivatives in any spacetime of any dimension to produce the components of the torsion and curvature. If the connection is symmetric the commutator vanishes, meaning that there is no torsion or curvature, and so a zero commutator implies flat spacetime in which the connection vanishes. A hypothetically symmetric connection is therefore zero. Therefore when applying the metric compatibility condition only antisymmetric connections can be used. The next section shows that if this rule is not followed, a self contradiction arises within Riemann geometry.

2. Metric compatibility and Evans identity

The most fundamental equation of cosmology is the commutator equation:

$$\left[D_\mu D_\nu\right]V^\rho = -T^\lambda_{\mu\nu}D_\lambda V^\rho + R^\rho_{\sigma\mu\nu}V^\sigma \tag{1}$$

in which the Riemann torsion:

$$T^\lambda_{\mu\nu} = \Gamma^\lambda_{\mu\nu} - \Gamma^\lambda_{\nu\mu} \tag{2}$$

and Riemann curvature:

$$R^\rho_{\sigma\mu\nu} = \partial_\mu\Gamma^\rho_{\nu\sigma} - \partial_\nu\Gamma^\rho_{\mu\sigma} + \Gamma^\rho_{\mu\lambda}\Gamma^\lambda_{\nu\sigma} - \Gamma^\rho_{\nu\lambda}\Gamma^\lambda_{\mu\sigma} \tag{3}$$

are deduced from the commutator of covariant derivatives. The covariant derivative of a vector V^σ is defined by:

$$D_\mu V^\sigma = \partial_\mu V^\sigma + \Gamma^\sigma_{\mu\lambda}V^\lambda \tag{4}$$

where $\Gamma^{\sigma}_{\mu\lambda}$ is the Christoffel connection. Both the torsion and the curvature are always antisymmetric in the indices of the commutator, μ and ν. Therefore, the connection is also antisymmetric in the same indices, μ and ν. If:

$$\mu = \nu \tag{5}$$

then the commutator, connection, torsion and curvature all vanish, in which case the spacetime becomes a spacetime that can be described by the ordinary four derivative. The commutator of such four derivatives is zero.

The metric compatibility equation [11] is:

$$D_{\rho} g_{\mu\nu} = \partial_{\rho} g_{\mu\nu} - \Gamma^{\lambda}_{\rho\mu} g_{\lambda\nu} - \Gamma^{\lambda}_{\rho\nu} g_{\mu\lambda} = 0 \tag{6}$$

If the connection is symmetric, it is zero, as just argued, so in this case the metric compatibility equation is:

$$\partial_{\rho} g_{\mu\nu} = 0 \tag{7}$$

and contains no connection. Therefore it is incorrect to use Eq. (6) with a symmetric connection, it can only be used with an antisymmetric connection.

The Evans identity [1–10] is:

$$D_{\mu} T^{\kappa\mu\nu} := R^{\kappa\mu\nu}_{\mu} \tag{8}$$

and is essentially the Hodge dual identity of the Cartan identity:

$$D_{\mu} \tilde{T}^{\kappa\mu\nu} := \tilde{R}^{\kappa\mu\nu}_{\mu} \tag{9}$$

where the tilde denotes Hodge duality. Therefore Eqs. (1), (6) and (8) are the fundamental equations of torsional cosmology. The Einstein field equation is incorrect and thoroughly obsolete because of its use of a symmetric connection. So in ECE type theory, the Einstein field equation is no longer used. The metrics are deduced from the orbital theorem of UFT 111 (www.aias.us). Given the antisymmetry of the connection therefore, the fundamental equations of torsional cosmology are:

$$D_{\rho} g_{\mu\nu} = 0 \tag{10}$$

and

$$D_{\mu} T^{\kappa\mu\nu} := R^{\kappa\mu\nu}_{\mu}. \tag{11}$$

As shown in the preceding paper UFT 186 (www.aias.us) all elements of the curvature tensor $R_{\mu}^{\kappa\mu\nu}$ vanish for the following metric:

$$g_{00} = 1 - \frac{r_0}{r} \tag{12}$$

$$g_{11} = -\left(1 - \frac{r_0}{r}\right)^{-1} \tag{13}$$

$$g_{22} = -r^2 \tag{14}$$

$$g_{33} = -r^2 \sin^2 \phi \tag{15}$$

where

$$r_0 = \frac{2MG}{c^2}. \tag{16}$$

Here M is the mass of an attracting object, G is Newton's constant, and c the vacuum speed of light. This metric should be regarded as an allowed solution of the orbital theorem of UFT 111, based on a spacetime of spherical symmetry. The appellation 'Schwarzschild metric' is incorrect, because in his only two extant papers of relevance (published in 1916) Schwarzschild did not derive the metric. What he actually did was to solve the Einstein field equation for two examples. The Einstein equation is now known to be flawed irretrievably because of its use of a symmetric connection, tantamount to neglect of torsion. The metric (12–16) is a fortuitously accurate description of a precessing elliptical orbit, and this lucky strike is what catalysed Einstein to fame, together with the flawed Eddington experiment. The metric fails completely for whirlpool galaxies, and the meaningless idea of dark matter introduced in a confused attempt to save the Einstein equation from obsolescence.

Written out in full, Eqs. (10) and (11) are:

$$\partial_\rho g_{\mu\nu} - \Gamma^{\lambda}_{\rho\mu} g_{\lambda\nu} - \Gamma^{\lambda}_{\rho\nu} g_{\mu\lambda} = 0 \tag{17}$$

and

$$\partial_\mu T^{\kappa}_{\nu\sigma} + \Gamma^{\kappa}_{\mu\lambda} T^{\lambda}_{\nu\sigma} - \Gamma^{\lambda}_{\mu\nu} T^{\kappa}_{\lambda\sigma} - \Gamma^{\lambda}_{\mu\sigma} T^{\kappa}_{\nu\lambda} = 0 \tag{18}$$

respectively using the rules [11] for the covariant derivative of a rank two and rank three tensor. The symmetries within these equations are:

$$g_{\mu\nu} = g_{\nu\mu} \tag{19}$$

$$T^{\kappa}_{\nu\sigma} = 2\Gamma^{\kappa}_{\nu\sigma} = -T^{\kappa}_{\sigma\nu}. \tag{20}$$

For a metric of type (12) the preceding paper UFT 186 (www.aias.us) shows that the use of metric compatibility with antisymmetric connection produces:

$$\Gamma^{0}_{10} = \frac{r_0}{2r(r-r_0)},\ \Gamma^{2}_{12} = \Gamma^{3}_{13} = \frac{1}{r},\ \Gamma^{3}_{23} = \frac{\cos\phi}{\sin\phi}. \tag{21}$$

If an erroneous attempt is made to use a symmetric connection in Eq. (17) results such as the following are obtained as in the preceding paper, UFT 186 (www.aias.us):

$$\Gamma^{1}_{11} = \frac{1}{2g_{11}} \partial_1 g_{11}. \tag{22}$$

This result gives a non-zero symmetric connection. This is not consistent with the commutator equation (1), which shows that all symmetric connections are zero. So the commutator equation must be used to define the antisymmetry of the connection, and the connection must be antisymmetric in the metric compatibility equation. The latter must be solved simultaneously with the Evans identity. Note that the commutator equation does not use the concept of metric at all, it uses only the connection, and thus defines the antisymmetry of the connection and the antisymmetry of torsion and curvature. The latter two tensors are the only two fundamental tensors of any spacetime in any dimension [11]. The Evans identity relates these two tensors, so is equally as fundamental as the commutator equation. The metric compatibility condition on the other hand asserts only that the covariant derivative of the metric vanishes and does not define curvature or torsion, nor does it define their antisymmetry. Using the metric compatibility condition without any other information, it is not possible therefore to define the antisymmetry of the connection. In logic therefore the antisymmetry of the connection can be defined only by the commutator equation, and so this antisymmetry property must be used in the metric compatibility equation.

If one tries to argue that the connection may have any symmetry (the obsolete point of view) then it may be decomposed into a sum of symmetric and anstisymmetric parts as follows:

$$\Gamma^{\lambda}_{\mu\nu} = \frac{1}{2}\left(\Gamma^{\lambda}_{\mu\nu} + \Gamma^{\lambda}_{\nu\mu}\right) + \frac{1}{2}\left(\Gamma^{\lambda}_{\mu\nu} - \Gamma^{\lambda}_{\nu\mu}\right) \tag{23}$$

using a well known theorem of matrices. If the connection is symmetric then:

$$\Gamma^{\lambda}_{\mu\nu} = \Gamma^{\lambda}_{\nu\mu}. \tag{24}$$

For example:

$$\mu = \nu = 1 \tag{25}$$

in which case:

$$[D_1, D_1]V^\rho = R^\rho_{\sigma 11} V^\sigma = 0 \tag{26}$$

and both the torsion and curvature vanish. This means that:

$$\left[D_\mu, D_\nu\right]V^\rho = \left[\partial_\mu, \partial_\nu\right]V^\rho = 0 \tag{27}$$

i.e. it means that the covariant derivative D_μ can be replaced by the ordinary derivative ∂_μ

The connection, however, is defined by [11]:

$$D_\mu V^\sigma = \partial_\mu V^\sigma + \Gamma^\sigma_{\mu\lambda} V^\lambda \tag{28}$$

so if

$$D_\mu V^\sigma = \partial_\mu V^\sigma \tag{29}$$

then

$$\Gamma^\sigma_{\mu\lambda} = 0. \tag{30}$$

The Christoffel connection vanishes if

$$\mu = \lambda \tag{31}$$

and so a symmetric connection is zero, Q. E. D.

Unfortunately this simple proof was not known in about 1900, when Levi-Civita proposed a symmetric connection. It was not known because the commutator equation was not known, and so the wrong connection symmetry was used uncritically for 110 years in a fog of dogma, or fogma, with catastrophic results for standard cosmology.

By use of computer algebra the following Riemann tensor elements were found to be non zero for the connections in Eq. (21):

$$R^0_{101} = \frac{-r_0\left(4r - 3r_0\right)}{4r^2\left(r - r_0\right)^2}, \tag{32}$$

$$R^{3}_{123} = -\frac{2\cos\phi}{r\sin\phi},\; R^{3}_{223} = 1.$$

Compatibility with the Evans identity means:

$$D_{\mu}T^{\kappa\mu\nu} = R^{\kappa\mu\nu}_{\mu}. \tag{33}$$

From Eq. (32) and using the methods of the preceding paper it is found that the identity reduces to:

$$-2\partial_{1}\Gamma^{0}_{10} = R^{010}_{1}. \tag{34}$$

The left hand side is:

$$-2\partial_{1}\Gamma^{0}_{10} = \frac{(2r - r_0)r_0}{r^2(r - r_0)^2} \tag{35}$$

and is not the same as the right hand side in general with the exception of the limit:

$$r \to \infty. \tag{36}$$

The so-called Schwarzschild metric is successful in the solar system only in this limit. This alone is enough to refute all claims to Big Bang and the existence of black holes.

The metric of a spherically symmetric spacetime [11] may be written as:

$$\left.\begin{aligned} g_{00} &= e^{2\alpha},\; g_{11} = -e^{2\beta}, \\ g_{22} &= -r^2,\; g_{33} = -r^2\sin^2\phi \end{aligned}\right]. \tag{37}$$

where α and β are in general functions of r and t. Computer algebra and hand calculations show that for such a metric the antisymmetric connections are:

$$\left.\begin{aligned} &\Gamma^{0}_{10} = \partial\alpha/\partial r,\; \Gamma^{1}_{01} = \partial\beta/\partial t, \\ &\Gamma^{2}_{12} = \Gamma^{3}_{13} = \frac{1}{r},\; \Gamma^{3}_{23} = \frac{\cos\phi}{\sin\phi}, \end{aligned}\right\} \tag{38}$$

and the non vanishing Riemann tensor elements are:

$$R^{0}_{001} = \frac{\partial\alpha}{\partial r}\frac{\partial\beta}{\partial t} + \frac{\partial^2\alpha}{\partial r\partial t}, \tag{39}$$

$$R^0_{101} = \frac{\partial^2 \alpha}{\partial r^2} + \left(\frac{\partial \alpha}{\partial r} \right)^2 , \tag{40}$$

$$R^1_{001} = -\left(\frac{\partial^2 \beta}{\partial t^2} + \left(\frac{\partial \beta}{\partial t} \right)^2 \right), \tag{41}$$

$$R^1_{101} = -\left(\frac{\partial \alpha}{\partial r} \frac{\partial \beta}{\partial t} + \frac{\partial^2 \beta}{\partial r \partial t} \right), \tag{42}$$

$$R^2_{012} = R^3_{013} = -R^3_{103} = \frac{1}{r} \frac{\partial \beta}{\partial t}, \tag{43}$$

$$R^3_{123} = -\frac{2 \cos \phi}{r \sin \phi}, \tag{44}$$

$$R^3_{223} = 1. \tag{45}$$

Applying the Evans identity as in note 187(10) accompanying this paper produces the constraint:

$$\frac{3\partial^2 \alpha}{\partial r^2} + \left(\frac{\partial \alpha}{\partial r} \right)^2 = 0 \tag{46}$$

and if this constraint is obeyed the metric does not violate the equations of torsional cosmology. This metric may therefore be an explanation of the dynamics of th esolar system and of an object such as the whirlpool galaxy provided that an appropriate function can be found. Note carefully that the metric does not depend on the use of the incorrect Einstein field equation.

References

[1] M. W. Evans, ed., J. Found. Phys. Chem., bimonthly from June 2011, published by Cambridge International Science Publishing (www.cisp-publishing.com).
[2] M .W. Evans, S. Crothers, H. Eckardt and K. Pendergast, Criticisms of the Einstein Field Equation (Cambridge International Science Publishing, 2011).
[3] K. Pendergast, The Life of Myron Evans (Cambridge International Science Publishing, 2011).
[4] The ECE websites, www.webarchive.org.uk, www.aias.us, www.atomicprecision.com, www.et3m.net, www.upitec.org.)
[5] M .W. Evans, H. Eckardt and D. W. Lindstrom, Generally Covariant Unified Field Theory (Abramis 2005 to 2011), in seven volumes.
[6] M .W. Evans, H. Eckardt and D. W. Lindstrom, ECE Theory Applied to Hydrogen Bonding (Serbian Academy of Sciences, 2010).
[7] M. W. Evans, ed., Modern Nonlinear Optics (Wiley, 2001, second edition), in three volumes; M. W. Evans and S. Kielich 9eds.), ibid, first edition, (Wiley 1992, 1993 and 1997), in three volumes.
[8] M .W. Evans and L. B. Crowell, Classical and Quantum Electrodynamics and the B(3) Field (World Scientific, 2001).
[9] M. W. Evans and J. - P. Vigier, The Enigmatic Photon (Kluwer, Dordrecht, 1994 to 2002), in ten volumes hardback / softback.
[10] M .W. Evans and A. A. Hasanein, The Photomagneton in Quantum Field Theory (World Scientific, 1994).
[11] S. P. Carroll, Spacetime and Geometry: and Introduction to General Relativity (Addison Wesley, New York, 2004).
[12] A very detailed criticism of the Einstein field equation is given in reference (1).

On the need for three metric compatibility equations to define a uniquely antisymmetric connection

M.W. Evans[1*], H. Eckardt[2] and D.W. Lindstrom[3*]**

**Alpha Institute for Advanced Studies (AIAS) (www.aias.us); **Unified Physics Institute of Technology (UPITEC) (www.upitec.org)*

It is shown that three metric compatibility equations are needed to define a unique antisymmetric connection. The need for an antisymmetric connection is indicated by the fundamental commutator method used to define the Riemann curvature and torsion. The use of one compatibility equation is not sufficient, and results in ambiguity.

Keywords: Unique antisymmetric connection, metric compatibility, ECE theory.

1. Introduction

Recently in this series of papers and books [1–10] developing ECE unified field theory it has been realized that the antisymmetric geometrical connection can be obtained straightforwardly from the metric compatibility condition and that the Hooke / Newton force equation of universal gravitation can be described directly in terms of the antisymmetric connection. The latter is defined at a most fundamental level by the commutator method [11] that is used to represent a round trip in the space with torsion and curvature. The Riemann curvature cannot be defined without the Riemann torsion. Both always occur together and ineluctably. In the early days of the subject, circa 1900, Levi Civita and Ricci defined the geometrical connection arbitrarily to be symmetric in its lower two indices and used this assertion to extract a unique connection from three compatibility equations in cyclic permutation. This was a catastrophic error repeated uncritically down the entire twentieth century. The commutator method shows clearly that the geometrical connection is antisymmetric in its lower two indices, it takes the

[1]e-mail: EMyrone@aol.com
[2]e-mail: horsteck@aol.com
[3]e-mail: dwlindstrom@gmail.com

antisymmetry of the commutator itself, as do the torsion and curvature tensors. A simple proof of this has been given in the preceding paper UFT 187 (www.aias.us). The commutator equation uses the concept of connection, and does not use the concept of metric. Similarly the most fundamental Evans identity [1–10] uses only the connection. The metric enters the stage because it is used for a variety of purposes [11] in geometry and it is used to derive orbits.

The metric compatibility equation is an assertion without proof and may be obtained from the tetrad postulate of Cartan's geometry [1–11], in which he was the first to generalize the idea of connection, due to Christoffel. The compatibility equation asserts that the covariant derivative of the metric vanishes. The metric is a rank two tensor so its covariant derivative contains two types of Christoffel connection. Using the idea of metric compatibility therefore, a link can be forged between the concept of metric, due to Riemann, and the concept of connection, due to Christoffel. The connection may be determined uniquely using three compatibility equations in cyclic permutation. It is shown in Section 2 that this procedure when used with a correctly antisymmetric connection produces a simple equation for the antisymmetric connection, and which defines the connection uniquely. In Section 3 computer algebra is used to show that the use of one compatibility equation is insufficient to define a connection, and leads to ambiguities.

2. Derivation of a uniquely defined antisymmetric connection

Consider three metric compatibility equations in cyclic permutation:

$$\partial_\rho g_{\mu\upsilon} - \Gamma^\lambda_{\rho\mu} g_{\lambda\upsilon} - \Gamma^\lambda_{\rho\upsilon} g_{\mu\lambda} = 0 \tag{1}$$

$$\partial_\mu g_{\upsilon\rho} - \Gamma^\lambda_{\mu\upsilon} g_{\lambda\rho} - \Gamma^\lambda_{\mu\rho} g_{\upsilon\lambda} = 0 \tag{2}$$

$$\partial_\upsilon g_{\rho\mu} - \Gamma^\lambda_{\upsilon\rho} g_{\lambda\mu} - \Gamma^\lambda_{\upsilon\mu} g_{\rho\lambda} = 0 \tag{3}$$

where $g_{\mu\upsilon}$ is the metric, and $\Gamma^\lambda_{\rho\mu}$ is the connection. Subtract the second two equations form the first to obtain:

$$\begin{aligned} &\partial_\rho g_{\mu\upsilon} - \partial_\mu g_{\upsilon\rho} - \partial_\upsilon g_{\rho\mu} - \Gamma^\lambda_{\rho\mu} g_{\lambda\upsilon} - \Gamma^\lambda_{\rho\upsilon} g_{\mu\lambda} \\ &+\Gamma^\lambda_{\mu\upsilon} g_{\lambda\rho} + \Gamma^\lambda_{\mu\rho} g_{\upsilon\lambda} + \Gamma^\lambda_{\upsilon\rho} g_{\lambda\mu} + \Gamma^\lambda_{\upsilon\mu} g_{\rho\lambda} = 0. \end{aligned} \tag{4}$$

From the commutator equation any non-zero connection must be antisymmetric:

$$\Gamma^\lambda_{\mu\upsilon} = -\Gamma^\lambda_{\upsilon\mu} \tag{5}$$

and the metric is defined to be symmetric [11]:

$$g_{\mu\upsilon} = g_{\upsilon\mu}. \tag{6}$$

Therefore:

$$\Gamma^{\lambda}_{\mu\upsilon} g_{\lambda\rho} + \Gamma^{\lambda}_{\upsilon\mu} g_{\rho\lambda} = 0. \tag{7}$$

For simplicity of development and without loss of generalization, consider a diagonal metric:

$$\mu = \upsilon \tag{8}$$

so

$$\Gamma^{\lambda}_{\mu\rho} g_{\upsilon\lambda} + \Gamma^{\lambda}_{\upsilon\rho} g_{\lambda\mu} = 0. \tag{9}$$

Therefore, Eq. (4) simplifies to:

$$\partial_{\rho} g_{\mu\upsilon} - \partial_{\mu} g_{\upsilon\rho} - \partial_{\upsilon} g_{\rho\mu} = \Gamma^{\lambda}_{\rho\mu} g_{\lambda\upsilon} + \Gamma^{\lambda}_{\rho\upsilon} g_{\mu\lambda}, \ \mu = \upsilon. \tag{10}$$

Let:

$$\mu = \upsilon = \alpha \tag{11}$$

then:

$$\partial_{\rho} g_{\alpha\alpha} - \partial_{\alpha} g_{\alpha\rho} - \partial_{\alpha} g_{\rho\alpha} = 2\Gamma^{\lambda}_{\rho\alpha} g_{\lambda\alpha}. \tag{12}$$

Multiply both sides by $g^{\lambda\alpha}$, the inverse metric, then:

$$g^{\lambda\alpha}\left(\partial_{\rho} g_{\alpha\alpha} - \partial_{\alpha} g_{\alpha\rho} - \partial_{\alpha} g_{\rho\alpha}\right) = 2g^{\lambda\alpha}\Gamma^{\lambda}_{\rho\alpha} g_{\lambda\alpha} = 2\Gamma^{\lambda}_{\rho\alpha} \tag{13}$$

Therefore:

$$\Gamma^{\lambda}_{\rho\alpha} = \frac{1}{2} g^{\lambda\alpha}\left(\partial_{\rho} g_{\alpha\alpha} - \partial_{\alpha} g_{\alpha\rho} - \partial_{\alpha} g_{\rho\alpha}\right). \tag{14}$$

The metric must be diagonal in this development, so:

$$\Gamma^{\alpha}_{\ \rho\alpha} = \frac{1}{2} g^{\alpha\alpha}\left(\partial_{\rho} g_{\alpha\alpha} - \partial_{\alpha} g_{\alpha\rho} - \partial_{\alpha} g_{\rho\alpha}\right) \tag{15}$$

because:

$$\alpha = \lambda. \tag{16}$$

The connection must be antisymmetric, so:

$$\rho \neq \alpha \tag{17}$$

in which case:

$$g_{\alpha\rho} = g_{\rho\alpha} = 0 \tag{18}$$

and

$$\Gamma^{\alpha}_{\rho\alpha} = \frac{1}{2} g^{\alpha\alpha} \partial_{\rho} g_{\alpha\alpha}. \tag{19}$$

This is a simple equation in which summation over repeated indices is not implied. Therefore for example, if

$$\alpha = 0, \rho = 1 \tag{20}$$

then

$$\Gamma^{0}_{10} = \frac{1}{2} \frac{1}{g_{00}} \partial_{1} g_{00}. \tag{21}$$

If

$$\alpha = 2, \ \rho = 1 \tag{22}$$

then:

$$\Gamma^{2}_{12} = \frac{1}{2} \frac{1}{g_{22}} \partial_{1} g_{22} \tag{23}$$

If

$$\alpha = 3, \rho = 1 \tag{24}$$

then

$$\Gamma^3_{13} = \frac{1}{2}\frac{1}{g_{33}}\partial_1 g_{33} \tag{25}$$

and if

$$\alpha = 3, \rho = 2 \tag{26}$$

then

$$\Gamma^3_{23} = \frac{1}{2}\frac{1}{g_{33}}\partial_2 g_{33} \tag{27}$$

Note that the index:

$$\alpha = 1 \tag{28}$$

is not used because this produces a symmetric connection, which is disallowed at the outset.

3. The non-vanishing antisymmetric connections of a sherically symmetric spacetime

As shown in the Appendix the use of three metric compatibility equations in cyclic permutation leads to the fundamental theorem (A11) linking the antisymmetric connection and the metric for any mathematical space of any dimension. The fundamental theorem is the metric compatibility condition for a connection defined to be antisymmetric. Conversely any antisymmetric metric is metric compatible by definition. The use of three metric compatibility equations leads back again to one metric compatibility equation. The latter is therefore necessary and sufficient to define the antisymmetric connection given a metric.

It is well known [11] that the metric of a spherically symmetric spacetime is:

$$g_{00} = -m(r,t), g_{11} = n(r,t), g_{22} = r^2,\ g_{33} = r^2 \sin^2 \phi \tag{29}$$

where $m(r,t)$ and $n(r,t)$ are functions of time t and the radial coordinate r. In this case the allowed antisymmetric connections are given in the following table.

For a metric of type (29) there are five antisymmetric connections defined as follows:

$$\Gamma^0_{10} = -\Gamma^0_{01} = \frac{1}{2m}\frac{\partial m}{\partial r} \tag{30}$$

$$\Gamma^1_{01} = -\Gamma^1_{10} = \frac{1}{2n}\frac{1}{c}\frac{\partial n}{\partial t} \tag{31}$$

$$\Gamma^2_{12} = -\Gamma^2_{21} = \frac{1}{r} \tag{32}$$

$$\Gamma^3_{13} = -\Gamma^3_{31} = \frac{1}{r} \tag{33}$$

$$\Gamma^3_{23} = -\Gamma^3_{32} = \frac{\cos\phi}{\sin\phi}. \tag{34}$$

These are constrained by the Evans identity [1–10]

$$D_\mu T^{\kappa\upsilon} = R^{\kappa}_{\mu}{}^{\mu\upsilon}. \tag{35}$$

The allowed elements of the torsion tensor for a spherical spacetime described by metric (29) are simply twice the allowed connection elements:

Table 1. The Antisymmetric Connections of a Spherical Spacetime

μ	υ	$\Gamma^\lambda_{\mu\upsilon} = -\Gamma^\lambda_{\upsilon\mu}$
0	1	Γ^0_{10}
	2	Γ^0_{20}
	3	Γ^0_{30}
1	0	Γ^1_{01}
	2	Γ^1_{21}
	3	Γ^1_{31}
2	0	Γ^2_{02}
	1	Γ^2_{12}
	3	Γ^2_{32}
3	0	Γ^3_{03}
	1	Γ^3_{13}
	2	Γ^3_{23}

$$T^{\lambda}_{\mu\upsilon} = 2\Gamma^{\lambda}_{\mu\upsilon}. \tag{36}$$

The curvature tensor is defined by

$$R^{\rho}_{\sigma\mu\upsilon} = \partial_{\mu}\Gamma^{\rho}_{\upsilon\sigma} - \partial_{\upsilon}\Gamma^{\rho}_{\mu\sigma} + \Gamma^{\rho}_{\mu\lambda}\Gamma^{\lambda}_{\upsilon\sigma} - \Gamma^{\rho}_{\upsilon\lambda}\Gamma^{\lambda}_{\mu\sigma}. \tag{37}$$

For use in Eq. (35) indices are raised and lowered with the metric elements as usual. With the collapse of Einsteinian cosmology, it is necessary to attempt the development of a new cosmology based on the antisymmetric connection. A spherical spacetime in which there is no dependence of r on t produces the metric (29) and the antisymmetric connections (30) to (34). The Evans identity (35) can be used to attempt to define m and n. Finally the equation of orbits is obtained from the infinitesimal line element:

$$ds^2 = -c^2 m(r,t)\,dt^2 + n(r,t)\,dr^2 + r^2 d\Omega^2 \tag{38}$$

The challenge is to describe all orbits without the use of dark matter. There are many types of orbits in astronomy, ranging from those in the solar system (the relativistic Kepler orbits) to those of spiral galaxies. It has been known for half a century that the obsolete Einsteinian cosmology was completely unable to describe the orbits of a spiral galaxy, so cosmology fell back on pure empiricism by introducing unobservable dark matter. At that stage cosmology ceased to be a scientific subject. The only way of putting it back on course is through use of the correct geometry as outlined in this and other papers of this series [1–10].

References

[1] M. W. Evans, ed., Journal of Foundations of Physics and Chemistry (bimonthly from June 2011, Cambridge International Science Publishing).

[2] M. W. Evans, S. Crothers, H. Eckardt and K. Pendergast, "Criticisms of the Einstein Field Equation" (Cambridge International Science Publishing, Spring 2011).

[3] K. Pendergast, "The Life of Myron Evans" (Cambridge International Science Publishing, Spring 2011).

[4] L. Felker, "The Evans Equations of Unified Field Theory" (Abramis 2007).

[5] M. W. Evans, H. Eckardt and K. Pendergast, "Generally Covariant Unified Field Theory" (Abramis 2005 - 2011) in seven volumes.

[6] The ECE websites: www.webarchive.org.uk, www.aias.us, www.atomicprecision.com, www.upitec.org, www.et3m.net.

[7] M. W. Evans (ed,), "Modern Nonlinear Optics" (Wiley, second edition, 2001) in three volumes; ibid., M. W. Evans and S.Kielich (eds.), (Wiley, first edition, 1992, 1993, 1997), in three volumes.

[8] M. W. Evans and L. B. Crowell, AClassical and Quantum Electrodynamics and the B(3) Field" (World Scientific, 2001).

[9] M. W. Evans and J.-P. Vigier, "The Enigmatic Photon" (Kluwer, 1994 - 2002), in ten volumes hardback and softback.

[10] M .W. Evans and A. A. Hasanein, "The Photomagneton in Quantum Field Theory" (World Scientific, 1994).

[11] S. P. Carroll, "Spacetime and Geometry: an Introduction to General Relativity" (Addison Wesley, New York, 2004).

Solar system orbits from the antisymmetric connection

M.W. Evans[1]*, H. Eckardt[2] and D.W. Lindstrom[3]***

**Alpha Institute for Advanced Studies (AIAS) (www.aias.us); **Unified Physics Institute of Technology (UPITEC) (www.upitec.org)*

The commutator equation, metric compatibility equation and Evans identity are used to derive straightforwardly a metric for solar system orbits. One solution for the metric is shown to have a closely similar r dependence to the empirical metric incorrectly known as the Schwarzschild metric. In general the metric is valid for spherically symmetric spacetime for any observed orbit, thus forging a new cosmology without the use of dark matter.

Keywords: ECE theory, antisymmetric metric, metric compatibility equation, Evans identity, solar system orbits, spherically symmetric spacetime.

1 Introduction

Recently in this series of papers [1-10] the fundamental theorem of Riemann geometry has been shown to be one based on an antisymmetric Christoffel connection. The fundamental theorem is the equation that derives the antisymmetric connection from the metric. In Section 2 it is shown that the metric must be redefined fundamentally for dimensional self consistency, the metric in the standard physics [11] is ill defined, so there are dimensional problems in the definition of the various connections. The new definition of the metric is self consistent and free of dimensional irregularities. In Section 3 It is shown that for a static metric (one independent of time), there is only one antisymmetric connection, a major step forward in the understanding of cosmology. This single connection must obey the Evans identity relating torsion and curvature, and this allows it to be deduced analytically. In Section 4 a computational and graphical analysis of the new metric is given, and it is shown that various solutions are possible, on of which is closely similar to the empirical attempt at understanding solar

[1]e-mail: EMyrone@aol.com

[2]e-mail: horsteck@aol.com

[3]e-mail: dwlindstrom@gmail.com

system orbits known misleadingly and incorrectly as the Schwarzschild metric.

2. Self consistent definition of the metric

With reference to note 189(1) accompanying this paper on www.aias.us, consider firstly the three dimensional Euclidean space [12]. Its metric elements are habitually defined as:

$$g_{11} = h_1^2,\ g_{22} = h_2^2,\ g_{33} = h_3^2 \tag{1}$$

where h_1, h_2 and h_3 are the scale factors. In the cylindrical polar system of coordinates

$$h_1 = 1,\ h_2 = r,\ h_3 = 1 \tag{2}$$

where r is the radial coordinate, so the habitual definition of the metric is:

$$g_{ij} = \begin{bmatrix} 1 & 0 & 0 \\ 0 & r^2 & 0 \\ 0 & 0 & 1 \end{bmatrix}. \tag{3}$$

In the Cartesian representation on the other hand:

$$g_{ij} = \begin{bmatrix} 1 & 0 & 0 \\ 0 & 1 & 0 \\ 0 & 0 & 1 \end{bmatrix}. \tag{4}$$

It is not consistent to have two different metrics representing the same mathematical space, and the presence of r^2 causes dimensional self inconsistency because r^2 has the units of square metres and the other metric elements in (3) are unitless. In consequence of the use of the habitual definition (3) extended to four dimensional Minkowski spacetime, some Christoffel symbols of the so called Schwarzschild metric do not vanish in a flat spacetime, in which all connections must be zero. This is a self contradiction that has gone uncorrected for a century.

The self inconsistency is eliminated by using the definition of the metric in terms of the unit vectors $\mathbf{e}_i$ of a curvilinear coordinate system, for example:

$$g_{ij} = \mathbf{e}_i \cdot \mathbf{e}_j \tag{5}$$

for a diagonal metric. So in both the Cartesian and cylindrical polar system, or any system of coordinates, the metric is always the unit diagonal and dimensionally self consistent:

$$g_{ij} = \begin{bmatrix} 1 & 0 & 0 \\ 0 & 1 & 0 \\ 0 & 0 & 1 \end{bmatrix}. \tag{6}$$

The Minkowski metric in any coordinate system is also coordinate free in the manner of Cartan's geometry:

$$g_{\mu\nu} = \begin{bmatrix} 1 & 0 & 0 & 0 \\ 0 & -1 & 0 & 0 \\ 0 & 0 & -1 & 0 \\ 0 & 0 & 0 & -1 \end{bmatrix}. \tag{7}$$

The metric of the spherical spacetime [11] is for all coordinate systems

$$g_{\mu\nu} = \begin{bmatrix} m(r) & 0 & 0 & 0 \\ 0 & -1/m(r) & 0 & 0 \\ 0 & 0 & -1 & 0 \\ 0 & 0 & 0 & -1 \end{bmatrix}. \tag{8}$$

In consequence there is only one antisymmetric connection in the spherically symmetric spacetime, a discovery of the utmost import. This connection is:

$$\Gamma^0_{10} = -\Gamma^0_{01} = \frac{1}{2m}\frac{\partial m}{\partial r} \tag{9}$$

and is constrained by the Evans identity [1–10]

$$D_\mu T^{\kappa\mu\nu} = R_\mu^{\kappa\mu\nu}. \tag{10}$$

In the habitually self inconsistent system the Minkowski metric was sometimes written in cylindrical polar coordinates as:

$$g_{\mu\nu} = \begin{bmatrix} 1 & 0 & 0 & 0 \\ 0 & -1 & 0 & 0 \\ 0 & 0 & -r^2 & 0 \\ 0 & 0 & 0 & -r^2 \sin^2\theta \end{bmatrix} \tag{11}$$

which is dimensionally incorrect and also contains the error:

$$h_3 = ?\ r \sin\theta \tag{12}$$

The correct expression for the third scale factor of the cylindrical polar system is well known [12] and should be:

$$h_3 = 1. \tag{13}$$

3 Derivation of the metric factor m(*r*)

The fundamental theorem of Riemann geometry for a diagonal metric of type (8) is [1–10]:

$$\Gamma^{\mu}_{\nu\mu} = \frac{1}{2g_{\mu\mu}} \partial_{\nu} g_{\mu\mu},\ \nu \neq \mu, \tag{14}$$

where there is no summation over the repeated μ index. The only connection that need be considered is

$$\Gamma^{0}_{10} = -\Gamma^{0}_{01}. \tag{15}$$

The torsion tensor is twice the connection:

$$T^{k\mu\nu} = 2\Gamma^{k\mu\nu} \tag{16}$$

and indices are raised and lowered as usual [11] using the metric and inverse metric:

$$\Gamma^{\kappa\mu\nu} = g^{\mu\alpha} g^{\nu\beta} \Gamma^{\kappa\mu\nu}_{\alpha\beta}\ . \tag{17}$$

From Eqs. (15) and (17):

$$\Gamma^{010} = g^{11} g^{00} \Gamma^{0}_{10}. \tag{18}$$

From Eqs. (10) and (15):

$$D_1 T^{010} = R_1^{010} \tag{19}$$

i.e.

$$D_1\left(g^{11} g^{00} T^{0}_{10}\right) = g^{11} g^{00} R^{0}_{110}. \tag{20}$$

By metric compatibility:

$$D_\mu g^{\nu\rho} = 0 \tag{21}$$

so

$$D_1\left(g^{11} g^{00}\right) - g^{11} D_1 g^{00} + g^{00} D_1 g^{11} - 0 \tag{22}$$

i.e.

$$D_1 T^{0}_{10} = R^{0}_{110}. \tag{23}$$

As in UFT 186 ff. on www.aias.us:

$$D_1 T^{0}_{10} = R^{0}_{110}. \tag{24}$$

so the Evans identity for the single connection (15) reduces to

$$\partial_1 T^{0}_{10} = R^{0}_{110}. \tag{25}$$

The curvature tensor is defined [1–11] as:

$$R^{\rho}_{\sigma\mu\nu} = \partial_\mu \Gamma^{\rho}_{\nu\sigma} - \partial_\nu \Gamma^{\rho}_{\mu\sigma} + \Gamma^{\rho}_{\mu\lambda}\Gamma^{\lambda}_{\nu\sigma} - \Gamma^{\rho}_{\nu\lambda}\Gamma^{\lambda}_{\mu\sigma} \tag{26}$$

and is known as the Riemann tensor, although it was not derived by Riemann because the idea of the geometrical connection was not known to Riemann.

The connection was first inferred by Christoffel in about 1867. The relevant curvature tensor is:

$$R^0_{110} = \partial_1 \Gamma^0_{01} - \partial_0 \Gamma^0_{11} + \Gamma^0_{1\lambda} \Gamma^\lambda_{01} - \Gamma^0_{0\lambda} \Gamma^\lambda_{11} \tag{27}$$

in which the following definition must hold:

$$\lambda = 0. \tag{28}$$

Therefore:

$$R^0_{110} = \partial_1 \Gamma^0_{01} + \Gamma^0_{10} \Gamma^0_{01}. \tag{29}$$

It follows that:

$$3\partial_1 \Gamma^0_{10} + \left(\Gamma^0_{10}\right)^2 = 0. \tag{30}$$

Using Eq. (9):

$$3\frac{\partial}{\partial r}\left(\frac{1}{m}\frac{\partial m}{\partial r}\right) + \frac{1}{2}\frac{1}{m^2}\left(\frac{\partial m}{\partial r}\right)^2 = 0 \tag{31}$$

which is a differential equation for m.
Use the change of variable:

$$m = \exp(2\alpha) \tag{32}$$

to obtain

$$\frac{3\partial^2 \alpha}{\partial r^2} + \frac{1}{2}\left(\frac{\partial \alpha}{\partial r}\right)^2 = 0. \tag{33}$$

Using the further change of variable:

$$\frac{1}{2}\left(\frac{\partial \alpha}{\partial r}\right)^2 = \frac{\partial \alpha}{\partial r} - f(r) \tag{34}$$

where f is a function of r, produces:

$$\frac{3\partial^2\alpha}{\partial r^2}+\frac{\partial\alpha}{\partial r}=f(r). \tag{35}$$

The reduced equation [13, 14] of Eq. (35) is:

$$\frac{3\partial^2\alpha}{\partial r^2}+\frac{\partial\alpha}{\partial r}=0. \tag{36}$$

The solution of the reduced equation (36) is the complementary function, and for dimensional correctness the complementary function must be:

$$\alpha=\pm\exp\left(\frac{-r}{3R}\right) \tag{37}$$

where R is a universal constant in units of metres. Therefore the complementary function is:

$$m_c(r)=\pm\exp\left(2\exp\left(\frac{-r}{3R}\right)\right) \tag{38}$$

and may be positive or negative valued. The particular integral of Eq. (35) is by definition the same as the particular integral of Eq. (33). The general solution of Eq. (33) is the sum of the particular integral and the complementary function [13, 14]. The general solution must be compared with experimental data in the solar system, these data are the orbits of planets, meteorites, and so forth, known to be precessing ellipses to an excellent approximation. In order to describe these data the following general solution is obtained:

$$m(r)=2-\exp\left(2\exp\left(\frac{-r}{3R}\right)\right) \tag{39}$$

in which the particular integral is:

$$m_p(r)=2 \tag{40}$$

and in which the complementary function is:

$$m_c(r) = -\exp\left(2\exp\left(\frac{-r}{R}\right)\right). \tag{41}$$

From purely empirical considerations it is known that a metrical function of the type:

$$m_{emp}(r) = 1 - \frac{r_0}{r} \tag{42}$$

produces solar system orbits very accurately. Here:

$$r_0 = \frac{2MG}{c^2} \tag{43}$$

where G is Newton's constant, M the mass of the sun and c the universal constant known as the vacuum speed of light. In the next section it is shown that the correctly deduced metrical function (39) can be approximated accurately by metric (42). The metric (39) is the first metric in cosmology that has been correctly derived from geometry. It is well known [1–10] that the older Einsteinian attempts incorrectly discarded torsion and for this reason are meaningless.

4 Computational analysis of the metric function

Details of cosmological solutions are presented in this section. After showing the general solution *m*(*r*) derived in the previous section, the metric function of Kepler orbits (relativistic and non-relativistic, i.e. Newtonian) is derived. These orbits are valid for the solar system. Finally we present the metric of logarithmic spiralling orbits being observed on galaxy scales.

4.1 Properties of the general solution

The general form of *m*(*r*) is given in Eq. (39). *R* is a constant and set to a numerical value of 1=3 for simplicity. From Fig. 1 it can be seen that this function behaves very similar to the so-called Scharzschild metric Eq. (42) with $r_0 = 1$. In particular there is a zero crossing and a divergent behaviour for $r \to 0$. The zero crossing appears at

$$r_{\text{zero}} = 3R\log\left(\frac{2}{\log(2)}\right). \tag{44}$$

The divergent behaviour can be avoided by shifting the r coordinate or using

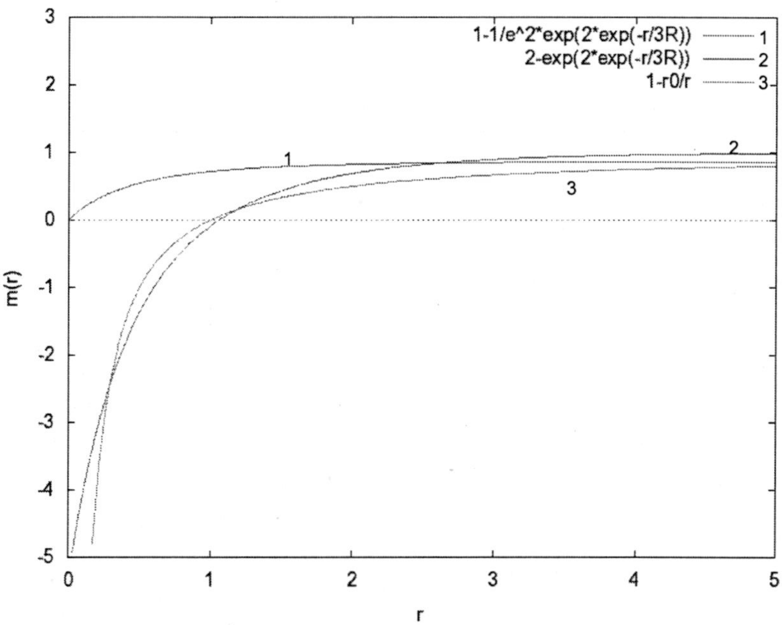

Fig. 1. Different forms of $m(r)$.

another admissible solution for the metric function, for example

$$m(r) = 1 - \frac{1}{e^2}\exp\left(2\exp\left(-\frac{r}{3R}\right)\right). \tag{45}$$

This function is regular for $r \geq 0$, however the limit of this function is

$$m(r) \to 1 - e^{-2} \tag{46}$$

instead of unity for r going to infinity.

From the solar system it is known that Eq. (42) gives an excellent descrition of gravitation. Therefore we tried to adopt the curve of $m(r)$ to the graph of this equation by least squares fitting. In $m(r)$ there is only one fitting parameter R available, therefore no perfect coincidence can be obtained. The numerical procedure gives

$$R = 0.374 r_0 \tag{47}$$

as an optimal value. The resulting graph is very similar to that shown in Fig. 1 for R = 1/3.

4.2 The relativistic and non-relativistic Kepler problem

The equation of orbits is found from Eq. (14) for $\mu = \nu = 0$:

$$\partial_0 g_{00} = 0 \tag{48}$$

which is

$$\frac{\partial}{\partial t} m(r,t) = 0. \tag{49}$$

The time dependence of m has not been considered so far. From Eq. (39) it is seen that only the characteristic radius R can be time dependent, therefore

$$m(r,t) = 2 - \exp\left(2\exp\left(-\frac{r}{R(t)}\right)\right). \tag{50}$$

Applying the time derivative in Eq. (49) then leads to the differential equation

$$\frac{1}{R^2(t)}\left(R(t)\frac{dr}{dt} - r\frac{dR(t)}{dt}\right) = 0 \tag{51}$$

or

$$\frac{dr}{dt} = \frac{r}{R(t)}\frac{dR(t)}{dt}. \tag{52}$$

This is an equation for all orbits. The radial coordinate r has to be considered to have a time dependence now which is characteristic for orbital motion. Note that this time dependence does not appear in the metric function (39) a priori.

The orbits of planets in the solar system are described experimentally by a precessing ellipse:

$$r = \frac{\alpha}{1 + \epsilon\cos(y\theta)} \tag{53}$$

with ϵ being the eccentricity, α the semi-major axis and y a parameter describing the precession of the ellipse. In the Newtonian limit we have

$$y \to 1. \tag{54}$$

The precessing ellipse is derived from the so-called Schwarzschild metric

$$ds^2 = \left(1-\frac{r_0}{r}\right)c^2dt^2 - \left(1-\frac{r_0}{r}\right)^{-1} dr^2 - r^2 \sin^2(\theta)d\theta^2 \tag{55}$$

which is an approximation to the metric with metric function $m(r, t)$ derived in this paper and passes into the Minkowski metric for

$$\frac{r_0}{r} \to 0. \tag{56}$$

The time derivative of θ in central motion is

$$\frac{d\theta}{dt} = \frac{L}{\mu r^2} \tag{57}$$

where L is the conserved angular momentum and μ the reduced mass. With (53) this is

$$\frac{d\theta}{dt} = \frac{L}{\mu\alpha^2}\left(1+\epsilon\cos(y\theta)\right)^2 \tag{58}$$

and from this equation and (53) follows by differentiation:

$$\frac{dr}{dt} = \frac{dr}{d\theta}\frac{d\theta}{dt} = \frac{Ly\epsilon}{\mu\alpha^2}\sin(y\theta). \tag{59}$$

Inserting the results into (52) gives a differential equation for $R(t)$:

$$\frac{1}{r}\frac{dr}{dt} = \frac{Ly\epsilon}{\mu\alpha^2}\left(1+\epsilon\cos(y\theta)\right)\sin(y\theta) = \frac{1}{R(t)}\frac{dR(t)}{dt}. \tag{60}$$

Because of

$$\frac{1}{R(t)}\frac{dR(t)}{dt} = \frac{d\log(R(t))}{dt} \tag{61}$$

this equation can be integrated to give

$$R(t) = c\frac{Ly\epsilon}{\mu\alpha^2}\int\left(1+\epsilon\cos(y\theta)\right)\sin(y\theta)\,dt. \tag{62}$$

with an integration constant c. The integral cannot be evaluated directly because of the time dependence of θ, but a variable substitution $t \to \theta$ can be performed using Eq. (58). Computer algebra then gives the final result

$$R(\theta) = \frac{c}{\left(1+\epsilon\cos(y\theta)\right)^{1/y}} \tag{63}$$

which with appropriate choice of the constant c can be written as

$$\boxed{R(\theta) = r(\theta)^{1/y}}. \tag{64}$$

For Newtonian orbits this further simplies to the fundamental result

$$\boxed{R(\theta) = r(\theta)}. \tag{65}$$

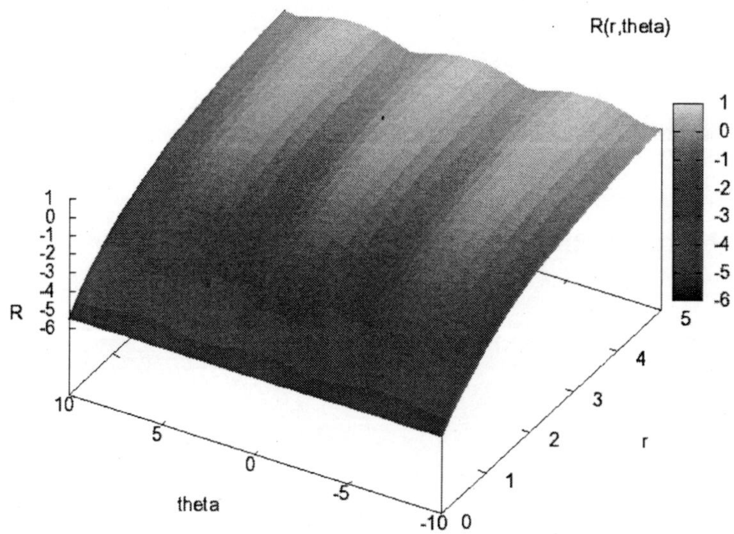

Fig. 2. Surface plot of $m(r, t)$ for Keplerian orbits.

The dependence of $m(r, \theta)$ has been graphed as a suface plot in Fig. 2 for $\epsilon = 0.3$ (all other constants set to unity). The cyclic weak dependence on the angle θ is visible. The dependence of θ from time can be calculated by integration of Eq. (58). The result obtained from computer algebra is quite complicated:

$$t = \frac{2\alpha^2\mu}{yL}\left(\frac{\operatorname{atan}\left(\frac{(2\epsilon-2)\sin(\theta y)}{2\sqrt{1-\epsilon^2}\,(\cos(\theta y)+1)}\right)}{2\sqrt{1-\epsilon^2}\,(\epsilon^2-1)} - \frac{\epsilon\sin(\theta y)}{(\cos(\theta y)+1)\left(\frac{(\epsilon^3-\epsilon^2-\epsilon+1)\sin(\theta y)^2}{(\cos(\theta y)+1)^2}-\epsilon^3-\epsilon^2+\epsilon+1\right)}\right). \tag{66}$$

The behaviour is illustrated in Fig. 3. It can be seen that for a relatively strong eccentricity of $\epsilon = 0.3$ the dependence remains near to linear as expected for the solar system.

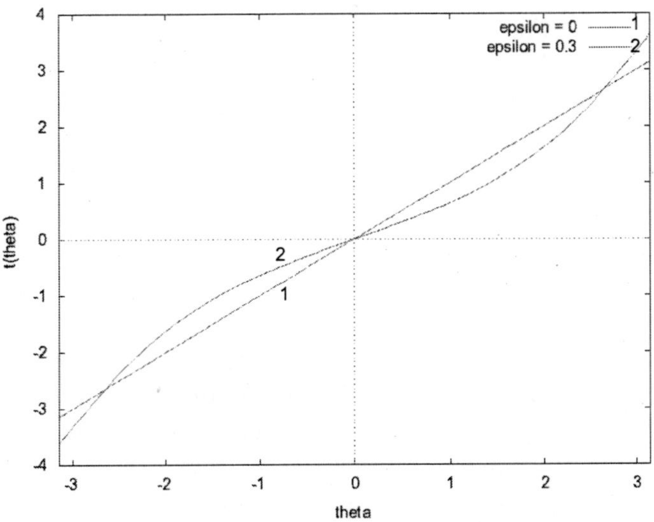

Fig. 3. Dependence $t(\theta)$ for Keplerian orbits.

4.3 Logarithmic spiral orbits

The orbit of a logarithmic spiral is given by [13]

$$r = k\exp(\alpha\theta) \tag{67}$$

with k and α being constants. The time dependence of r is

$$r(t) = \left(\frac{2\alpha L}{\mu}t + k^2C\right)^{1/2} \tag{68}$$

where μ and L are dened as for the Keplerian orbits. From the latter equation follows

$$\frac{dr(t)}{dt} = -\frac{\alpha L}{\mu}\left(\frac{2\alpha L}{\mu}t + k^2C\right)^{-3/2}. \tag{69}$$

With Eq. (52), which is

$$\frac{1}{r}\frac{dr}{dt} = \frac{1}{R(t)}\frac{dR(t)}{dt}, \tag{70}$$

we obtain the the differential equation

$$-\frac{\alpha L}{\mu}\left(\frac{2\alpha L}{\mu}t + k^2C\right)^{-2} = \frac{1}{R(t)}\frac{dR(t)}{dt}. \tag{71}$$

The solution is

$$R(t) = R_0\exp\left(-\frac{1}{\mu^2}\left(\alpha^2L^2t^2 + \alpha k^2\mu CLt\right)\right) \tag{72}$$

with an integration constant R_0. This is a Gaussian function, depicted in Fig. 4 where all constants have been set to unity again. Inserting $R(t)$ into Eq. (50) gives the metric function for logarithmic orbits. This is also plotted in Fig. 4, showing some kind of inverse Gaussian behaviour. The combined r and t dependence can be observed in the surface plot (Fig. 5). $m(r, t)$ behaves like

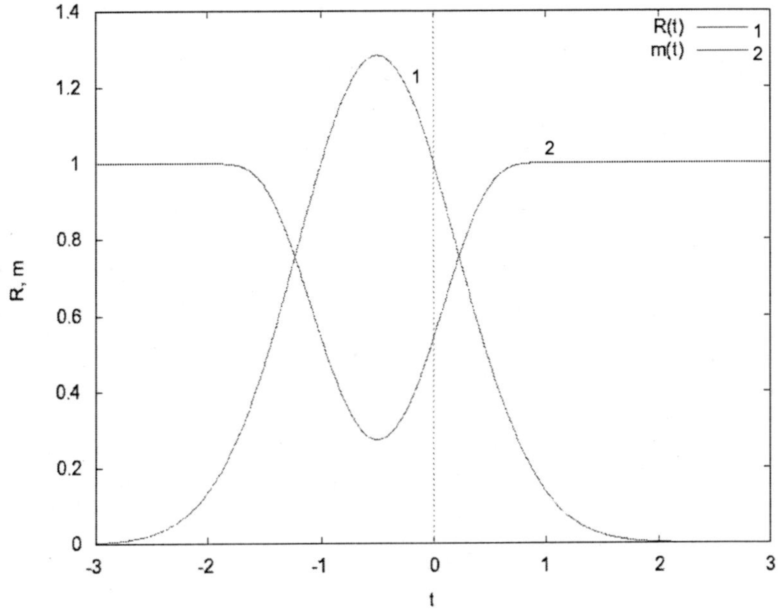

Fig. 4. Functions $R(t)$ and $m(r, t)$ (with $r = 5$) for a logarithmic spiral orbit.

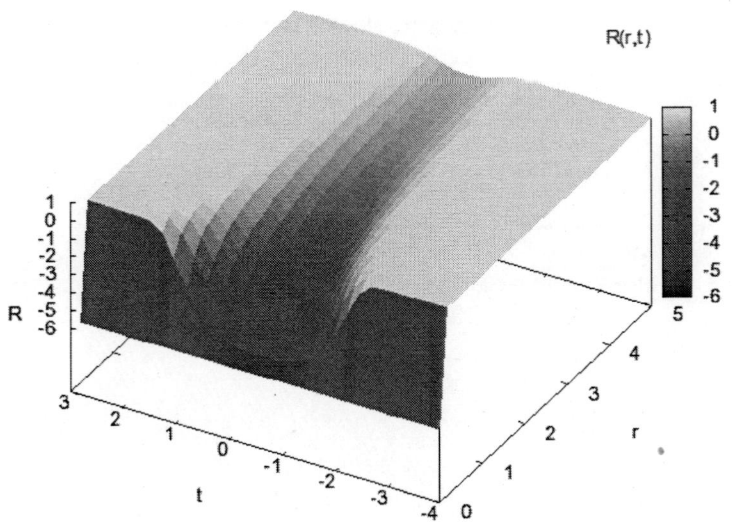

Fig. 5. Surface plot of $m(r$, t) for a logarithmic spiral orbit.

a Gaussian in time and an $1/r$ function in the radial coordinate. The physically meaningful range begins at $t = 0$ where m has the highest slope, reflecting the fact that m deviates most from free-space behaviour near to the the center of a spiral. Spiral arms of galaxies should be describable in this way.

References

[1] M. W. Evans, Ed., Journal of Foundations of Physics and Chemistry (Cambridge International Science Publishing, bimonthly from June 2011, www.cisp-publishing.com).

[2] M. W. Evans, S. Crothers, H. Eckardt and K. Pendergast, "Criticisms of the Einstein Field Equation" (Cambridge International Science Publishing, Spring 2011).

[3] K. Pendergast, "The Life of Myron Evans" (Cambridge International Science Publsishing, Spring 2011).

[4] M .W. Evans, H. Eckardt and D. W. Lindstrom, "Generally Covariant Unified Field Theory" (Abramis Academic, 2005 to 2011), in seven volumes.

[5] L. Felker, "The Evans Equations of Unified Field Theory" (Abramis Academic, 2007, available on www.aias.us in English and Spanish).

[6] M .W. Evans, D. W. Lindstrom and H. Eckardt, "ECE Theory Applied to H Bonding" (Serbian Academy of Sciences, 2010).

[7] The ECE open source websites: www.webarchive.org.uk, www.aias.us, www.atomicprecision.com, www.upitec.org, www.et3m.net

[8] M. W. Evans and S. Kielich, (Eds.), "Modern Nonlinear Optics" (Wiley, 1992, 1993, 1997, 2001), in six volumes and two editions.

[9] M. W. Evans and L. B. Crowell, "Classical and Quantum Electrodynamics and the B(3) Field" (World Scientific 2001).

[10] M .W. Evans and J.-P. Vigier, "The Enigmatic Photon" (Kluwer 1994 to 2002, hardback and softback), in ten volumes; M. W. Evans and A. A. Hasanein, "The Photomagneton in Quantum Field Theory" (World Scientific, 1994).

[11] S. P. Carroll, "Spacetime and Geometry: an Introduction to General Relativity" (Addison Wesley, New York, 2004).

[12] E. G. Milewski (Chief Ed.), "The Vector Analysis Problem Solver" (Res. Ed. Assoc., New York, 1987).

[13] J. B. Marion and S. T. Thornton, "Classical Dynamics of Particles and Systems" (Harcourt Brace, New York, 1988, 3rd. ed.).

[14] G. Stephenson, "Mathematical Methods for Science Students" (Longmans, 1968, 5th impression).

Disintegration of Einsteinian general relativity, towards a new ECE cosmology

M.W. Evans[1*] and H. Eckardt[2]**

**Alpha Institute for Advanced Studies (AIAS) (www.aias.us); **Unified Physics Institute of Technology (UPITEC) (www.upitec.org)*

It is shown by a combination of incisive analysis and computational methods that the Einsteinian general relativity is severely self-inconsistent. In order to replace it completely, a new cosmology is suggested based on the correct antisymmetry of the Christoffel connection and metric compatibility subject to the Evans identity of differential geometry. The disintegration of Einsteinian relativity begs the question of whether the claims to precision in tests of the theory can ever be meaningful, the basic theory being riddled with errors of concept and mathematics.

Keywords: ECE theory, new cosmology, disintegration of Einsteinian general relativity.

1. Introduction

In recent papers of this series of 191 papers to date [1–10], the beginnings of a new cosmology have been forged using the antisymmetric Christoffel connection, the metric compatibility equation and Evans identity of differential geometry. The results have been expressed in terms of the metric of spherical spacetime [11] and a function m of the radial coordinate r of the cylindrical polar system. This metric has been used to describe an orbit in a plane. In this paper it is shown that the Einsteinian general relativity is severely self inconsistent and essentially disintegrates under the mildest scholarly scrutiny. This fragility of Einstein's general relativity has been known for more than ninety years, but has been ignored for obscure, non-scientific, reasons. In UFT 150 and 155 of this series, and in various essays and broadcasts on www.aias.us,, the Einsteinian theory of light deflection and time delay is heavily criticised. In the monograph of reference (2), 'Criticisms of the Einstein Field Equation', all metrics of the

[1]e-mail: EMyrone@aol.com
[2]e-mail: horsteck@aol.com

equation are shown to be incorrect computationally due to neglect of torsion. Crothers in that monograph [2[gives a review of his own and other criticisms of the Einstein field equation. In ref. [12], Einstein's calculation of the precession of the ellipse is heavily criticised. Less than a month after this paper appeared in late 1915, K. Schwarzschild [12[pointed out basic errors in it in a letter to Einstein in which Schwarzschild proposed a metric. This does not take the format of the so called 'Schwarzschild metric' used in 'standard' physics, now an entirely obsolete subject. This is pointed out most clearly by Crothers [2] and in ref. [12], in which it is pointed out that this paper on precession is essentially the only one written by Einstein on the topic.

In Section 2 it is shown straightforwardly that the so called Schwarzschild solution does not give an ellipse in general, and that it does not reduce correctly to the Newtonian limit. So called 'precision tests' of the Einstein theory are therefore meaningless. This has been known since the late nineteen fifties through the velocity curve of a spiral galaxy, but again facts have been ignored in favour of dogma and 'dark matter'. In contrast the new ECE cosmology based on the m function is able to describe solar system and galactic orbits with the correct geometry and symmetry. In Section 3 a computational analysis is made of some of Einstein's claims, and the claims of those who adhere to the 'Schwarzschild dogma'. It is shown that they fail on every count.

2. Mathematics in place of dogma

Consider an orbit in a plane defined by:

$$dZ = 0. \tag{1}$$

In the spherically symmetric spacetime the infinitesimal line element in cylindrical polar coordinates is:

$$ds^2 = c^2 d\tau^2 = m(r) c^2 dt^2 - m^{-1}(r) dr^2 - r^2 d\theta^2 \tag{2}$$

where in general m can be a function of r and t [11]. Here τ is the proper time and m the mass of an object in orbit, such as a planet. In this section the function m is considered to be a function only of r. The Hamiltonian is:

$$H = \frac{1}{2} mc^2 = \frac{1}{2} m \left(m(r) \left(\frac{dt}{d\tau} \right)^2 - \frac{1}{m(r)} \left(\frac{dr}{d\tau} \right)^2 - r^2 \left(\frac{d\theta}{d\tau} \right)^2 \right). \tag{3}$$

The constants of motion [1–10] are the total energy:

$$E = m(r) mc^2 \frac{dt}{d\tau}, \tag{4}$$

the angular momentum

$$L = mr^2 \frac{d\theta}{d\tau}, \tag{5}$$

and the linear momentum

$$p = \frac{m}{m(r)} \frac{dr}{d\tau}. \tag{6}$$

Therefore, Eq. (3) is:

$$\frac{1}{2} m \left(\frac{dr}{d\tau} \right)^2 = \frac{1}{2} \left(\frac{E^2}{mc^2} - m(r) \left(mc^2 + \frac{L^2}{mr^2} \right) \right). \tag{7}$$

The orbital equation is obtained from:

$$\frac{dr}{d\tau} = \frac{dr}{d\theta} \frac{d\theta}{d\tau} = \frac{L}{mr^2} \frac{dr}{d\theta}. \tag{8}$$

From Eqs. (7) and (8):

$$\frac{1}{m} \left(\frac{L}{r^2} \right)^2 \left(\frac{dr}{d\theta} \right)^2 = \frac{E^2}{mc^2} - m(r) \left(mc^2 + \frac{L^2}{mr^2}. \right) \tag{9}$$

so:

$$\left(\frac{dr}{d\theta} \right)^2 = r^4 \left(\frac{1}{b^2} - m(r) \left(\frac{1}{a^2} + \frac{1}{r^2} \right) \right) \tag{10}$$

where the constants a and b have the units of metres and are defined by:

$$a = \frac{L}{mc}, \quad b = \frac{Lc}{E}. \tag{11}$$

Therefore:

$$\frac{dr}{d\theta} = r^2 \left(\frac{1}{b^2} - m(r) \left(\frac{1}{a^2} + \frac{1}{r^2} \right) \right)^{1/2} \tag{12}$$

for any spherically symmetric spacetime.

The tangential velocity of the planet is obtained from:

$$v = \frac{dr}{dt} = \frac{dr}{d\tau}\frac{d\tau}{dt} = \frac{dr}{d\theta}\frac{d\theta}{d\tau}\frac{d\tau}{dt} \tag{13}$$

with:

$$\frac{d\theta}{d\tau} = \frac{L}{mr^2}, \frac{d\tau}{dt} = \frac{m(r)}{E}mc^2. \tag{14}$$

So

$$v = cbm(r)\left(\frac{1}{b^2} - m(r)\left(\frac{1}{a^2} + \frac{1}{r^2}\right)\right)^{1/2} \tag{15}$$

and again this is true for any spherically symmetric spacetime.

The angular velocity of the orbit is obtained from:

$$\omega = \frac{d\theta}{dt} = \frac{d\theta}{dr}\frac{dr}{dt} = cb\frac{m(r)}{r^2} \tag{16}$$

and this is the simplest test of general relativity. Yet this test seems never to have been applied experimentally.

The so called 'standard model' of physics uses a function:

$$m(r) = 1 - \frac{r_0}{r} \tag{17}$$

where:

$$r_0 = \frac{2MG}{c^2}. \tag{18}$$

Here G is Newton's constant, c the speed of light and M the mass of the attracting object. In the solar system M is the mass of the Sun. In his letter to Einstein of Dec. 1915 [12], Schwarzschild proposed the function:

$$m(r) = 1 - \frac{\gamma}{\left(R^3 + r_0^3\right)^{1/3}} \tag{19}$$

where R is not Schwarzschild's function (19) does not have a singularity, unlike the function (17). It is unknown who introduced function (17) and fabricated its source historically.

It is claimed [11] that the function (17) produces a precessing ellipse:

$$r = \frac{\alpha}{1+\epsilon\cos(x\theta)} \tag{20}$$

where 2α is the latus rectum, ϵ the eccentricity and x a constant. The methods underlying this claim are given in detail in note 190(9) on www.aias.us. They are dubious because they rely on the self contradiction of 'effective potential' from Eq. (12) used in the classical or Newtonian description, and also rely on dubious approximations explained in note 190(9). Finally, planetary data [11] do not verify the theory. The precession of the ellipse is a particularly difficult piece of data to choose because it is influenced by other factors well known in astronomy. Einstein [12] wrote only one paper on this topic and then abandoned it. In order to test this claim correctly the following equation must be used (note 190(7)):

$$\frac{dr}{d\theta} = \frac{x\alpha\epsilon\sin(x\theta)}{(1+\epsilon\cos(x\theta))^2} = r^2\left(\frac{1}{b^2} - m(r)\left(\frac{1}{a^2}+\frac{1}{r^2}\right)\right)^{1/2} \tag{21}$$

so the m function needed for a precessing ellipse is:

$$m(r) = \frac{a^2}{r^2\left(a^2+r^2\right)}\left(\frac{r^4}{b^2} - \left(\frac{x\alpha\epsilon}{(1+\epsilon\cos(x\theta))^2}\right)^2 \sin^2(x\omega)\right) \tag{22}$$

where

$$\cos(x\theta) = \frac{1}{\epsilon}\left(\frac{\alpha}{r} - 1\right). \tag{23}$$

It is shown using Maxima (computer algebra) in Section 3 that: in general this function is not the function (17). So the Einstein field equation does not produce a precessing ellipse in general. However, the function (22) can be described by suitable parameterization of the fundamental m function [1–10] of ECE cosmology:

$$m(r) = 2 - \exp\left(2\exp\left(-\frac{r}{R}\right)\right). \tag{24}$$

It is not known whether Einstein himself ever used the function (17), but Eq. (17) saturates the standard dogma as an archetypical idol of the cave. Writing out Eq. (15) it becomes:

$$v = \frac{mc^2}{E}\left(1-\frac{r_0}{r}\right)\left(\frac{E^2}{m^2c^2} - c^2 - \frac{L^2}{m^2r^2} + \frac{2MG}{r} + \frac{2MGL^2}{m^2c^2r^3}\right)^{1/2}. \tag{25}$$

It is further claimed that this function reduces to Newton=s:

$$v=\left(\frac{2E}{m}+\frac{2MG}{r}-\frac{L^2}{m^2r^2}\right)^{1/2} \tag{26}$$

where in Eq. (26), the total energy is:

$$E=\frac{1}{2}mv^2+\frac{L^2}{2mr^2}-\frac{MG}{r}. \tag{27}$$

Even without computer algebra it is seen that Eq. (25) never reduces to Eq. (27) for the following reasons.
1) It has to be assumed that:

$$\frac{mc^2}{E}\left(1-\frac{r_0}{r}\right)=1. \tag{28}$$

2) It has to be assumed that:

$$\frac{E^2}{m^2c^2}-c^2=\frac{2E}{m} \tag{29}$$

3) If it is assumed that in Eq. (28):

$$E=mc^2, 1-\frac{r_0}{r}=1, \tag{30}$$

then it must also be assumed that:

$$v=cb\left(\frac{1}{b^2}-\frac{1}{a^2}-\frac{1}{r^2}\right) \tag{31}$$

and this never gives Eq. (26).
4) From Eq. (29):

$$E^2-2Emc^2-m^2c^4=0 \tag{32}$$

i.e.

$$E=\left(1\pm\sqrt{2}\right)mc^2 \tag{33}$$

and this contradicts Eq. (30).

The dogmatists claim that

$$E = mc^2\left(1 - \frac{r_0}{r}\right)\frac{dt}{d\tau} \tag{34}$$

in the metric:

$$c^2 d\tau^2 = c^2 dt^2\left(1 - \frac{r_0}{r}\right) - \left(1 - \frac{r_0}{r}\right)^{-1} dr^2 - r^2 d\theta^2. \tag{35}$$

By definition:

$$c^2 d\tau^2 = c^2 dt^2\left(1 - \frac{r_0}{r}\right) - d\mathbf{r}\cdot d\mathbf{r} \tag{36}$$

where:

$$d\mathbf{r}\cdot d\mathbf{r} = v^2 dt^2. \tag{37}$$

So:

$$c^2 d\tau^2 = c^2 dt^2\left(1 - \frac{r_0}{r}\right) - v^2 dt^2 \tag{38}$$

and:

$$\frac{dt}{d\tau} = \left(1 - \frac{r_0}{r} - \frac{v^2}{c^2}\right)^{-1/2} \tag{39}$$

The total energy is therefore:

$$E = mc^2\left(1 - \frac{r_0}{r}\right)\left(1 - \frac{r_0}{r} - \frac{v^2}{c^2}\right)^{-1/2} = \text{constant}. \tag{40}$$

This may be approximated by:

$$\frac{E}{mc^2} \sim \left(1 - \frac{r_0}{r}\right)\left(1 + \frac{1}{2}\left(\frac{r_0}{r} + \left(\frac{v}{c}\right)^2\right)\right) \tag{41}$$

if:

$$\frac{r_0}{r} \ll 1, \quad \frac{v}{c} \ll 1. \tag{42}$$

However, from Eq. (30)

$$\frac{E}{mc^2} = 1 \tag{43}$$

in which case:

$$\frac{r_0}{r} \to 0 \tag{44}$$

and

$$\frac{E}{mc^2} \to 1 + \frac{1}{2}\left(\frac{v}{c}\right)^2. \tag{45}$$

This result is usually interpreted as the relativistic kinetic energy of special relativity for a free particle m and no attractive potential. The procedure of trying to force Eq. (40) to become Eq. (27) means that mc^2 must be subtracted from E, and that the potential of attraction must vanish. In this case however the contradiction (31) is present. Comparing Eqs (45) and (30) the velocity vanishes! The Schwarzschild dogma cannot give the Newtonian limit, contrary to saturation propaganda.

Numerical methods as in Section 3 show this in another way by using

$$v = cbm(r)\left(\frac{1}{b^2} - m(r)\left(\frac{1}{a^2} + \frac{1}{r^2}\right)\right)^{1/2} = \left(\frac{2E}{m} + \frac{2MG}{r} - \frac{L^2}{m^2 r^2}\right)^{1/2} \tag{46}$$

in which case the Newtonian limit can be obtained from the dogmatists if and only if the following cubic is solved for m , and then if and only if the result is Eq. (17). The actual result from computer algebra is an intricate function of r that never becomes Eq. (17). Even worse for the dogmatists is that the m function from Eq. (46) is not the same as the m function from Eq. (22), and neither is Eq. (17). So the Einsteinian general relativity disintegrates under the mildest and most obvious scholarly scrutiny.

The only way forward is to adopt the correct ECE theory in a series of new experiments designed to produce a self-consistent function $m(r)$. By Ockham's Razor the simplest experiment or astronomical observation is that of the angular velocity:

$$\omega = \frac{d\theta}{dt} = \frac{cbm(r)}{r^2} \tag{47}$$

by measuring the displacement of angle θ in time t directly, and by measuring the distance r directly. Contemporary methods can be used with great precision. These astronomical observations give m (r) directly. If the orbital linear velocity of for example a planet can be measured unequivocally without hidden assumptions, then:

$$v = cbm(r)f(r), f(r) = \left(\frac{1}{b^2} - m(r)\left(\frac{1}{a^2} + \frac{1}{r^2}\right)\right)^{1/2} \tag{48}$$

and the function m should be the same, experimentally, as found from Eq. (47). If this is not true then the spacetime is not spherically symmetric, or general relativity is not valid. This test requires knowledge of the total energy E and angular momentum L. Finally, if the orbit can be measured unequivocally without hidden assumptions, the same m should be found from:

$$\frac{dr}{d\theta} = r^2 f(r). \tag{49}$$

As shown in detail in note 190(15), (www.aias.us), there are some dubious assumptions in the textbook approach to Kepler's second law (equal areas in equal times). Astronomical measurements are all inter-related, so great care must be taken to determine whether some hidden assumption is present in them. The optimal method would be to measure quantities as directly as possible. Having determined the m function experimentally, it can be fitted to the fundamental function (24) given by geometry. Computer algebra must be used to find the correct method of reaching the Newtonian limit, which must be a limit of the m function. In the case of angular velocity for a circular orbit, Eq. (47) shows that the Newtonian limit is:

$$\omega = \frac{cb}{r^2} \tag{50}$$

The reader is referred to background notes 190(1) to 190(15) on www.aias.us for more details.

Finally in this Section the problem of the whirlpool galaxy is considered by using the observation that the stars are arranged in a logarithmic spiral:

$$r = r_0 \exp(\alpha\theta) \tag{51}$$

of high pitch α, which means that the outer arms become straight lines as observed directly in astronomy. The whirlpool galaxy is characterized by its well known velocity curve, in which the orbital velocity v becomes constant as the radial coordinate r becomes infinite. This is completely non Einsteinian and non Newtonian. From Eq. (51):

$$\frac{dr}{d\theta} = \alpha r \tag{52}$$

and from Eqs. (52) and (12):

$$f(r) = \frac{\alpha}{r}. \tag{53}$$

Therefore the orbital velocity is:

$$v = \frac{cb\alpha}{r}\left(\frac{r^2}{b^2} - \alpha^2\right)\left(\frac{a^2}{r^2 + a^2}\right). \tag{54}$$

Astronomical data show that as:

$$r \to \infty,\ v \to v_\infty = \text{constant}. \tag{55}$$

From Eq. (54):

$$v = cb\alpha\left(\frac{r}{b^2} - \frac{\alpha^2}{r}\right)\left(\frac{a^2}{r^2 + a^2}\right). \tag{56}$$

i.e.

$$v = \frac{ca^2}{b}\alpha\left(\frac{r}{r^2 + a^2}\right) - \frac{cb\alpha^3}{r}\left(\frac{a^2}{r^2 + a^2}\right). \tag{57}$$

In the limit:

$$r \to \infty,\ \alpha \to \infty \tag{58}$$

so:

$$\lim_{r \to \infty} \frac{a^2}{r^2 + a^2} \to 0. \tag{59}$$

Therefore:

$$v \to v_\infty = \frac{ca^2}{b}\alpha\left(\frac{r}{r^2+a^2}\right). \tag{60}$$

Considering the limit:

$$\lim_{r\to\infty} \frac{r}{r^2+a^2} = \lim_{r\to\infty} \frac{1}{r}\left(1+\frac{a^2}{r^2}\right)^{-1} \to \frac{1}{r} \tag{61}$$

the limiting velocity is:

$$v_\infty \to \frac{ca^2}{b}\left(\frac{\alpha}{r}\right) \tag{62}$$

and is a constant if:

$$\frac{\alpha}{r} \to \text{constant.} \tag{63}$$

The dogma fails for the whirlpool galaxy because it asserts that the function (17) gives the Newtonian limit, in which case the velocity curve reaches a maximum and falls back again to zero, in complete contradiction of the data, where the velocity curve reaches a plateau.

3. Numerical analysis with Maxima

We give some examples for the nature of the function m(r) in various contexts showing that inconsistencies for this function arise. There is no transition from Einsteinian theory to Newtonian mechanics.

The angular change or radius for an orbiting body is in Einsteinian as well as ECE theory given by

$$\frac{dr}{d\theta} = r^2\left(\frac{1}{b^2} - m(r)\left(\frac{1}{a^2}+\frac{1}{r^2}\right)\right)^{1/2} \tag{64}$$

where the functions $m(r)$ are given by

$$m(r) = 1 - \frac{r_0}{r} \tag{65}$$

in Einsteinian and

$$m(r) = 2 - \exp\left(2\exp\left(-\frac{r}{R}\right)\right) \tag{66}$$

in ECE theory with constants of motion

$$a = \frac{L}{mc},\ b = \frac{Lc}{E}, \tag{67}$$

see Eqs. (11), (12), (17) and (24). The simplest transition to Newtonian theory would be

$$m(r) \rightarrow 1\,. \tag{68}$$

Then, Eq. (65) becomes

$$\frac{dr}{d\theta} \rightarrow r^2\left(\frac{1}{b^2} - \frac{1}{a^2} - \frac{1}{r^2}\right)^{1/2}. \tag{69}$$

Because of (68) it is

$$\frac{a}{b} = \frac{E}{mc^2}. \tag{70}$$

Since E consists of the rest energy plus kinetic energy, we have

$$E > mc^2, \tag{71}$$

hence

$$a > b \text{ and } \frac{1}{a^2} < \frac{1}{b^2} \tag{72}$$

but the difference is small because of the huge rest energy of a celestial body. Consequently the condition from (69) for obtaining a real value of the square root,

$$\frac{1}{b^2} \geq \frac{1}{a^2} + \frac{1}{r^2}, \tag{73}$$

can barely be fulfilled for $a \approx b$. The condition is much easier to fulfill for $m(r) < 1$, which is guaranteed for all values of r. This can be seen from the graphs of Fig. 1 and 2. In Fig. 1 there are bound states for the m functions of Einsteinian and ECE theory which are compatible with elliptic orbits. However, in order to obtain a non-negative square root argument for $m = 1$, one has to choose $a \gg b$. This is typical for the non-relativistic case. In other words, the

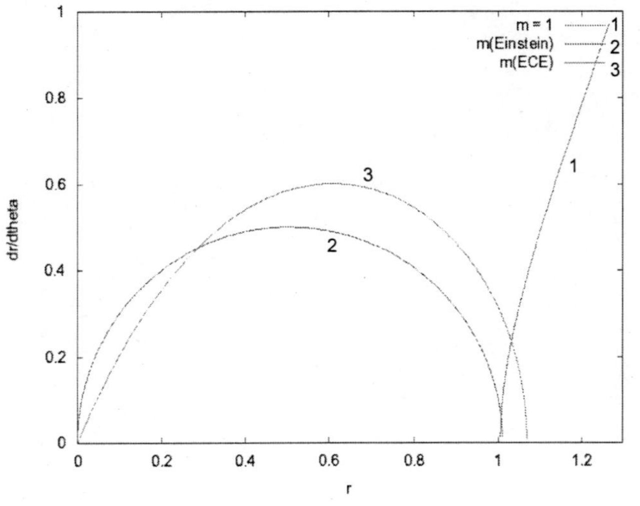

Fig. 1. Bound states of $dr/d\theta$, for functions m (Einstein) and m (ECE). Parameters were $a = b=10$, $R = 1$, $r_0=1$. For comparison: $m=1$ with $a = 10$, $b = 1$.

function $m = 1$ can only be used in the non-relativistic case where the r-dependent m function is defined for total energies with exclusion of the rest energy. This is one reason why both types of theory do not pass into one another. In Fig. 2 there is no bound state for m (Einstein) and two distinct regions, one bound and one unbound, for m (ECE). Both cases do not describe ellipses.

If it is assumed that $dr/d\theta$ describes an orbits of a precessing ellipse, the general form of orbit has to be equated by $dr/d\theta$ of elliptic orbits which has been done in Eq. (21). As a result, a particular form of function $m(r)$ is obtained, see (22). The radial part of this function is graphed in Fig. 3 for three values of eccentricity ϵ. It can be seen that these curves differ only in the region very near to the centre (note that we have set $r_0 = 1$ throughout our calculations). However these curves differ significantly from the general form of m presented in Eq. (66). This may be the reason why it was very difficult to find bound elliptic orbits in Figs. 1 and 2.

The angular dependence of m is shown in Fig. 4 in a polar diagram. For small radii, there is a significant angular variation. For radii larger than r_0 these differences become indistinguishable as can also be seen from Fig. 3. This behaviour is to be expected for a spherical spacetime, there should be no angular variance, otherwise the geometry would not be spherical. The behaviour near to the center probably plays no practical role since this is inside a star where additional laws of physics are valid.

In section 2 an expression for the radial velocity of an orbiting body was derived (Eq. (15)). In the non-relativistic approximation this expression should shade into the Newtonian expression given in Eq. (26). From equating both velocity expressions (Eq. (46)), one should obtain the function m which represents

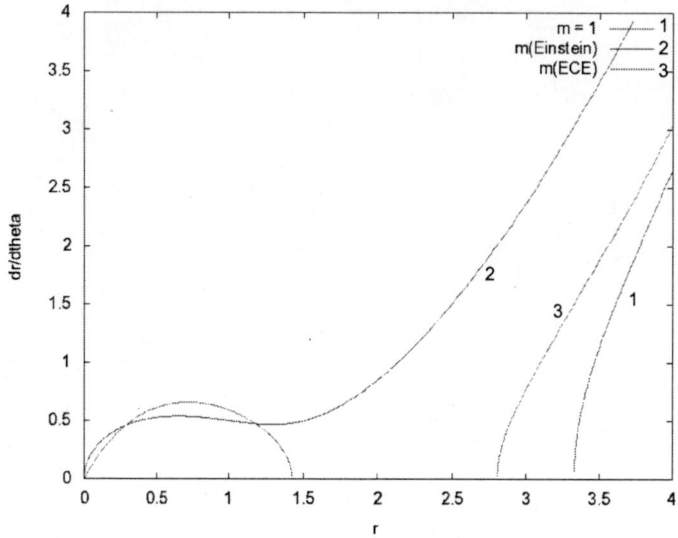

Fig. 2. Unbound states of $dr/d\theta$, for functions m =1, m (Einstein) and m (ECE). Parameters were a = 2.5, b = 2, R = 1, r_0 = 1.

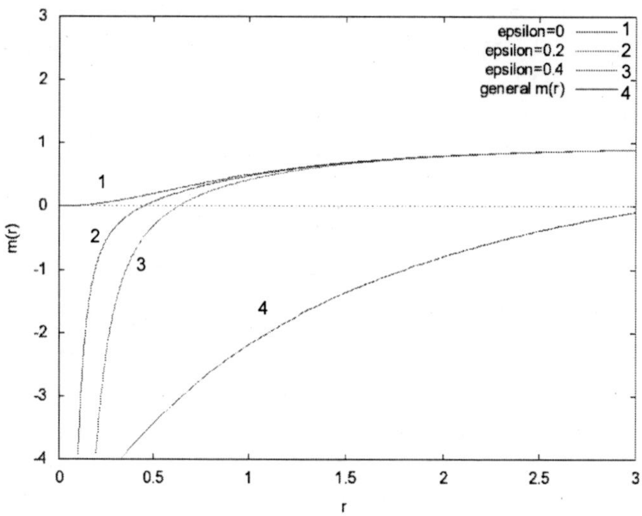

Fig. 3. $m(r)$ of Eq. (22) for $a = b = x = 1$, $\theta = \pi/2$, in comparison with $m(r)$ from Eq. (24).

the Newtonian theory. Due to the different energy definitions we do not expect a result of $m = 1$. Squaring Eq. (46) gives

$$c^2b^2m^2(r)\left(\frac{1}{b^2}-m(r)\left(\frac{1}{a^2}+\frac{1}{r^2}\right)\right)=\frac{2E}{m}+\frac{2MG}{r}-\frac{L^2}{m^2r^2}. \tag{74}$$

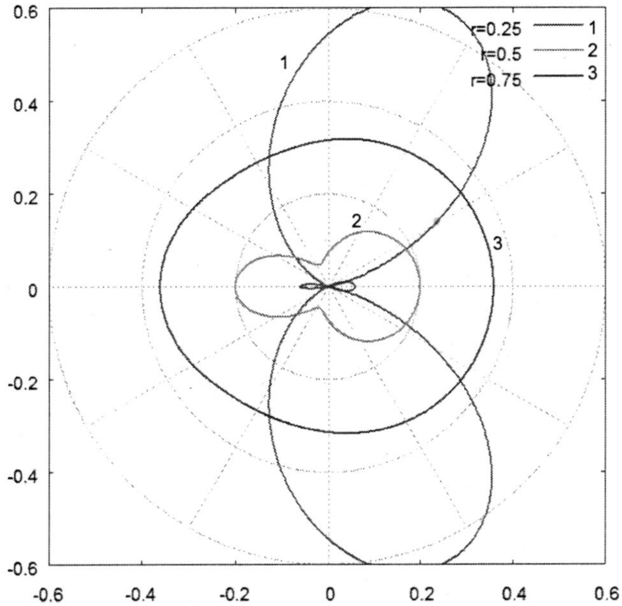

Fig. 4. Angular dependence of $m(r)$ from Eq. (22) for three radius values, $\epsilon = 0.2$.

This is a cubic equation in $m(r)$ of the form

$$c_1 m^3(r) + c_2 m^2(r) + c_3 = 0 \tag{75}$$

with constants

$$c_1 = -c^2 b^2 \left(\frac{1}{a^2} + \frac{1}{r^2} \right), \tag{76}$$

$$c_2 = c^2, \tag{77}$$

$$c_3 = -\frac{2E}{m} - \frac{2MG}{r} + \frac{L^2}{m^2 r^2}. \tag{78}$$

For consistency reasons, we write the constants L and E in terms of a and b as obtained from Eq.(67)

$$L = amc, \; E = \frac{a}{b} mc^2. \tag{79}$$

Equation (75) can be solved, giving two complex and one real solution. The real solution is

$$m(r)=\left(\frac{\sqrt{c_3\left(27c_1^2c_3+4c_2^3\right)}-\frac{27c_1^2c_3+2c_2^3}{54c_1^3}}{2 3^{\frac{3}{2}}c_1^2}\right)^{1/3}+$$
$$+\frac{c_2^2}{9c_1^2\left(\frac{\sqrt{c_3\left(27c_1^2c_3+4c_2^3\right)}}{2 3^{\frac{3}{2}}c_1^2}-\frac{27c_1^2c_3+2c_2^3}{54c_1^3}\right)^{1/3}}-\frac{c_2}{3c_1}. \tag{80}$$

Graphing this highly complicated expression yields the results shown in Fig. 5. All tested parameter combinations give qualitatively the same curve type. The limit of m for large r is negative and not +1 as it should be. In particular this is not the function used by Einstein theory. This result makes evident that Newtonian and Einstein theory are not compatible.

Finally, we investigated the form of $m(r)$ for a whirlpool galaxy. Equating both terms for $dr/d\theta$ as before, Eqs. (12) and (52), we obtain

$$r^2\left(\frac{1}{b^2}-m(r)\left(\frac{1}{a^2}+\frac{1}{r^2}\right)\right)^{1/2}=\alpha r \tag{81}$$

or

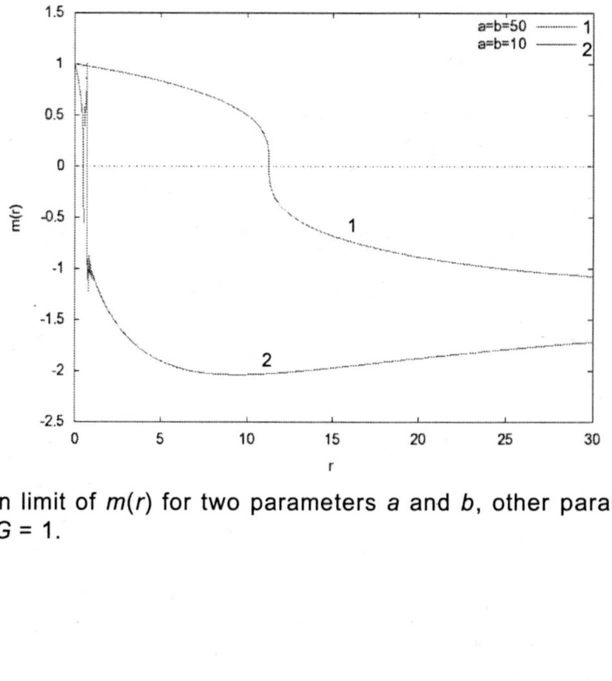

Fig. 5. Newtonian limit of $m(r)$ for two parameters a and b, other parameters being c = 1, m = 1, M = 100, G = 1.

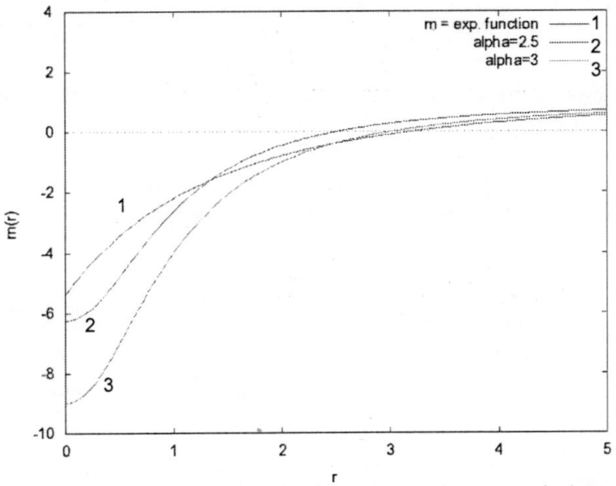

Fig. 6. $m(r)$ for a spiral galaxy, with parameters $a = 1$, $b = 1$.

$$m(r) = \frac{a^2}{r^2 + b^2}\left(\frac{r^2}{b^2} - \alpha^2\right) \tag{82}$$

which has been graphed in Fig. 6. The curve is similar to the general m function, Eq. (66), for values of α between 2.5 and 3. The far field limit is unity as required but convergence behaviour is slightly different for all three curves. Compared to Fig. 3, this indicates that descibing spiral galaxies by ECE theory may even be simpler than describing elliptical orbits, a surprising result when taking into account that dark matter had to be assumed to bring Einstein theory in agreement with the experimental velocity curve. In total we have shown by numerical methods that the Einstein theory is inconsistent and untenable.

References

[1] M .W. Evans, H. Eckardt and D. Lindstrom, "Generally Covariant Unified Field Theory" (Abramis, 2005 - 2011), in seven volumes.

[2] M. W. Evans, S. Crothers, H. Eckardt and K. Pendergast, "Criticisms of the Einstein Field Equation" (Cambridge International Science Publishing, Spring 2011).

[3] Kerry Pendergast, "The Life of Myron Evans" (Cambridge International Science Publishing, Spring 2011).

[4] M .W. Evans, Ed., J. Foundations of Physics and Chemistry, (Cambridge International Science Publishing, from June 2011, bimonthly).

[5] L. Felker, "The Evans Equations of Unified Field Theory" (Abramis 2007, Spanish translation on www.aias.us).

[6] The ECE websites: www.aias.us and weblog (both archived on www.webarchive.org.uk), www.atomicprecision.com, www.upitec.org, www.et3m.net.

[7] M. W. Evans, H. Eckardt and D. W. Lindstrom, Proc. Serbian Academy of Sciences (2010).

[8] M .W. Evans, Ed., "Modern Nonlinear Optics" (Wiley 2001, second edition) in three volumes; M. W. Evans and Kielich (Eds.), ibid., first edition (1992, 1993, 1997) in three volumes.

[9] M . W. Evans and L. B. Crowell, "Classical and Quantum Electrodynamics and the B(3) Field"

(World Scientific, 2001).

[10] M .W. Evans and J.-P. Vigier, "The Enigmatic Photon" (Kluwer, 1994 to 2002 hardback and softback), in ten volumes.

[11] S. P. Carroll, "Spacetime and Geometry: an Introduction to General Relativity" (Addison Wesley, New York, 2004).

[12] A. A. Vankov, www.wbabin.net/eeuro/vankov.pdf, , A. Einstein, Proc. Royal Prussian Acad., 831 - 839 (22nd Nov., 1915); original letter by K. Schwarzschild to Einstein dated 22nd. Dec., 1915, both original source documents reproduced by Vankov.

Towards an ECE cosmology: Precession of the perihelion, whirlpool galaxies and binary pulsars

M.W. Evans[1*], H. Eckardt[2] and R. Cheshire***

**Alpha Institute for Advanced Studies (AIAS) (www.aias.us); **Unified Physics Institute of Technology (UPITEC) (www.upitec.org)*

A new cosmology is developed based on spherically symmetric spacetime and the use of one antisymmetric connection in accordance with ECE theory. The new cosmology is illustrated with a precessing ellipse (solar system), whirlpool galaxy and binary pulsar and a self consistent description of all systems given in terms of spherically symmetric spacetime described by an infinitesimal line element characterized by a function m of the radial coordinate r. Using this method it is shown straightforwardly that the standard model description of the solar system is wildly erroneous, and that claims to precision tests of this erroneous theory cannot be true. The new ECE cosmology is analyzed with computer algebra.

Keywords: ECE cosmology, solar system, whirlpool galaxy, binary pulsar.

1. Introduction

In recent papers of this series of 192 papers to date [1–10] on Einstein Cartan Evans (ECE) unified field theory a new cosmology has been initiated using the antisymmetric connection, metric compatibility and Evans identity. In UFT 190 (www.aias.us) severe problems were encountered in the claims by the standard model of physics, notably, the Schwarzschild metric is incorrectly attributed to him. It is well known that the claims by Einstein are riddled with errors [11, 12] and what appear to be deliberate obfuscations, so Einstein's role in physics has been revised negatively by contemporary scholarship. In particular Schwarzschild [11] strongly criticised Einstein's calculation of the precession of the perihelion in a letter of 22nd Dec. 1915. Subsequently criticisms of this calculation have multiplied [12]. During the course of development of ECE theory [1–12] many more errors have been uncovered in Einsteinian general relativity, the most

[1]e-mail: EMyrone@aol.com
[2]e-mail: horsteck@aol.com

serious being the neglect of torsion, the use of an incorrect symmetry for the Christoffel connection. This means that the Einstein field equation is incorrect, as is well known by now, and all metrics and cosmologies based on that equation are incorrect and should be discarded as meaningless dogma.

In Section 2 a new cosmology is suggested based on a spherical spacetime characterized by an infinitesimal line element in a plane and a function *m* of the radial coordinate *r*. The orbital equation is deduced straightforwardly from this line element and compared with the analytical equation of the observed orbit. In the solar system this has been observed since ancient times to be a precessing ellipse. So m for the solar system can be derived by elementary calculus checked by computer algebra. Using this method it is easy to show that the standard model is complete nonsense, a classic example of Langmuir's pathological science or repeated dogma. The Newtonian limit is defined by using the correct *m*, whose characteristics are analysed by computer. This method is applied to the whirlpool galaxy, in which the stars are arranged on a logarithmic spiral. So the *m* function of a whirlpool galaxy is easily obtained by elementary calculus again checked by computer. It is well known that Einsteinian general relativity fails completely for a whirlpool galaxy. Finally the method is extended to the binary pulsar, in which two objects orbit in a precessing ellipse which slowly spirals inwards. Using the m function the correct method of calculating the precession of the perihelion is given in the solar system.

In Section 3 the results of Section 2 are analysed with computer algebra and by schematics that clarify the orbits visually.

2. The *m* function for the solar system, whirlpool galaxy and binary pulsar

Consider the infinitesimal line element of a spherically symmetric spacetime:

$$ds^2 = c^2 d\tau^2 = m(r)c^2 dt^2 - \frac{dr^2}{m(r)} - r^2 d\theta^2 \tag{1}$$

in the plane:

$$dZ = 0 \tag{2}$$

in the cylindrical polar coordinates (r, θ). Here τ is the proper time and m is a function characteristic of a spherically symmetric spacetime. In recent papers of this series, m has been derived from ECE theory using a single antisymmetric Christoffel connection. One possible result of this theory is:

$$m(r) = 2 - \exp(2\exp(-r/R)) \tag{3}$$

where R is a characteristic distance. This result assumes that m is a function only of r. This paper derives m for various observed orbits and fits them by computer to Eq. (3). For a precessing ellipse:

$$r=\frac{\alpha}{1+\epsilon\cos(x\theta)} \tag{4}$$

where alpha is a basic property of the ellipse, its semi-right magnitude with units of metres. Here ϵ is the eccentricity and x the precession constant. Therefore:

$$\frac{dr}{d\theta}=\frac{x\epsilon}{\alpha}r^2\sin(x\theta). \tag{5}$$

However, the orbital equation can also be derived from Eq. (1) [1–10] using well known methods and is:

$$\frac{dr}{d\theta}=r^2\left(\frac{1}{b^2}-m(r)\left(\frac{1}{a^2}+\frac{1}{r^2}\right)\right)^{1/2} \tag{6}$$

where a and b are constants of motion with the units of metres defined by:

$$a=\frac{L}{mc},\ b=\frac{Lc}{E}. \tag{7}$$

Here E is the total energy and L the total angular momentum, both of which are constants of motion, and m (not to be confused with $m(r)$) is the mass of the attracted object (a planet for example). So:

$$\left(\frac{1}{b^2}-m(r)\left(\frac{1}{a^2}+\frac{1}{r^2}\right)\right)^{1/2}=\frac{x\epsilon}{\alpha}\sin(x\theta) \tag{8}$$

and the correct m function in the solar system is therefore:

$$m(r)=\left(\frac{1}{b^2}-\left(\frac{x\epsilon}{\alpha}\right)^2+\left(\frac{x}{\alpha}\right)^2\left(1-\frac{\alpha}{r}\right)^2\right)\left(\frac{1}{a^2}+\frac{1}{r^2}\right)^{-1} \tag{9}$$

where we have used:

$$\cos(x\theta)=\frac{1}{\epsilon}\left(\frac{\alpha}{r}-1\right). \tag{10}$$

The standard model uses the incorrect function:

$$m(r)=?\ 1-\frac{r_0}{r} \tag{11}$$

in which:

$$r_0=\frac{2MG}{c^2}. \tag{12}$$

Here G is Newton's constant, M is the mass of the attracting object and c the vacuum speed of light. The function (11) is incorrectly attributed to K. Schwarzschild. This attribution is not true. On 22nd Dec. 1915 Schwarzschild proposed [11]:

$$m(R)=1-\frac{\gamma}{\left(R^3+r_0^3\right)^{1/3}} \tag{13}$$

where γ is a constant and R is different from r. Clearly, the function (11) is not the same as the function (9) obtained directly from a precessing ellipse. This is enough to show that the standard model is nonsense. Another way of showing this is by using Eq. (11) in Eq. (6):

$$ds^2=c^2d\tau^2=?\left(1-\frac{r_0}{r}\right)c^2dt^2-\left(1-\frac{r_0}{r}\right)^{-1}dr^2-r^2d\theta^2. \tag{14}$$

Comparison with Eq. (5) gives:

$$\frac{x\epsilon}{\alpha}\sin(x\theta)=\left(\frac{1}{b^2}-\left(1-\frac{r_0}{r}\right)\left(\frac{1}{a^2}+\frac{1}{r^2}\right)\right)^{1/2} \tag{15}$$

which is a polynomial in r. The roots of this polynomial are constants, a result which has no meaning because it implies that Eq. (11) is true only at the roots of the polynomial (15), and not otherwise. On a logical plane it is difficult to see why such arrant nonsense has lasted for more than ninety years. The claim to 'precision testing' of such nonsense is deception on a large scale, and Einstein himself is suspected [11, 12] of obfuscation.

In the limit:

$$r\to\infty \tag{16}$$

the m function behaves as follows:

$$m(r)\underset{r\to\infty}{\to} a^2\left(\frac{1}{b^2}-\left(\frac{x\epsilon}{\alpha}\right)^2+\left(\frac{x}{\alpha}\right)^2\right). \tag{17}$$

For the planet Earth for example the observed perihelion precession is:

$$5.0\pm1.2 \text{ arcsecond/century} \tag{18}$$

In a revolution of 2π radians (a year):

$$x\theta \to x\theta+2\pi x. \tag{19}$$

If an initial measurement is made at some point in the orbit, the earth advances by:

$$x\theta \to x\theta+\left(2\pi+\frac{0.05}{3600}\right) \tag{20}$$

in radians. So:

$$2\pi x=2\pi+\frac{0.05}{3600} \tag{21}$$

and:

$$x-1=2.21\times10^{-6}. \tag{22}$$

Therefore, x is very close to unity. The eccentricity of the Earth's orbit is:

$$\epsilon=0.01671123 \tag{23}$$

Its aphelion and perihelion are respectively:

$$\left.\begin{aligned} r_{\max}&=1.52098252\times10^{12}\,\text{m},\\ r_{\min}&=1.47098290\times10^{12}\,\text{m}.\end{aligned}\right\} \tag{24}$$

Therefore:

$$\alpha=(1+\epsilon)r_{\min}=(1-\epsilon)r_{\max}=1.4956\times10^{12}\,\text{m}. \tag{25}$$

So to an excellent approximation:

$$m(r)=\left(\frac{E}{mc^2}\right)^2\left(\frac{1}{1+\left(\frac{L}{mcr}\right)^2}\right) \tag{26}$$

because:

$$\frac{x\epsilon}{\alpha}\ll 1,\ \frac{x}{\alpha}\ll 1,\ x=1. \tag{27}$$

The earth's orbit is nearly circular so to an excellent approximation:

$$L=mr^2\omega \tag{28}$$

where the angular velocity is

$$\omega=\frac{d\theta}{dt} \tag{29}$$

so m is a simple function:

$$m(r)=\left(\frac{E}{mc^2}\right)^2\left(\frac{1}{1+\left(\frac{\omega r}{c}\right)^2}\right). \tag{30}$$

Obviously this is not Eq. (11). In Section 3 the characteristics of the correct solar system m function are evaluated by computer.

The standard model of physics must be discarded in favour of science. For example the following method is suggested for calculating x, effectively the precession of the perihelion. The semiminor and semimajor axes of the ellipse are defined respectively by:

$$r_1=\frac{\alpha}{\left(1-\epsilon^2\right)^{1/2}},\ r_2=\frac{\alpha}{1-\epsilon^2}. \tag{31}$$

From Eq. (4):

$$\Delta\theta=\theta_2-\theta_1=\frac{1}{x}\left[\cos^{-1}\left(\frac{\alpha}{r_2}-1\right)-\cos^{-1}\left(\frac{\alpha}{r_1}-1\right)\right] \tag{32}$$

and from Eq. (6):

$$\Delta\theta = \int_{r_1}^{r_2} \frac{1}{r^2}\left(\frac{1}{b^2} - m(r)\left(\frac{1}{a^2} + \frac{1}{r^2}\right)\right)^{-1/2} dr. \tag{33}$$

Therefore:

$$\cos^{-1}\left(\frac{\alpha}{r_2} - 1\right) - \cos^{-1}\left(\frac{\alpha}{r_1} - 1\right) = x\int_{r_1}^{r_2} \frac{1}{r^2}\left(\frac{1}{b^2} - m(r)\left(\frac{1}{a^2} + \frac{1}{r^2}\right)\right) dr \tag{34}$$

in which:

$$m(r) = \left(\frac{1}{b^2} - \left(\frac{x\epsilon}{\alpha}\right)^2 + \left(\frac{x}{\alpha}\right)^2\left(1 - \frac{\alpha}{r}\right)^2\right)\left(\frac{1}{a^2} + \frac{1}{r^2}\right)^{-1}. \tag{35}$$

The distances r_1 and r_2 have been observed since ancient times, so x can be calculated from Eq. (34) by computer to machine precision. It should be the same as the experimental value (18). If not then general relativity itself fails or the spacetime cannot be considered to be spherically symmetric. As can be seen from an inspection of Eq. (34) the method based on the perihelion precession is a very poor one in the solar system. It is far simpler to deduce m from the angular velocity [1–10]:

$$\omega = \frac{d\theta}{dt} = \left(\frac{Lc^2}{E}\right)\frac{m(r)}{r^2}. \tag{36}$$

Although much simpler than perihelion precession, this method seems never to have been used.

In a whirlpool galaxy the stars are observed to be distributed in a logarithmic spiral:

$$r = r_0 \exp(\zeta\theta) \tag{37}$$

in which ζ is the pitch (UFT 190 on www.aias.us)). It has been known for half a century that the Einsteinian general relativity fails completely to describe the velocity curve of a whirlpool galaxy. From Eq. (37):

$$\frac{dr}{d\theta} = \zeta r \tag{38}$$

so

$$m(r) = \left(\frac{1}{b^2} - \frac{\zeta^2}{r^2}\right)\left(\frac{1}{a^2} + \frac{1}{r^2}\right)^{-1} \tag{39}$$

which is a simple well behaved function with no singularity graphed in Section

3. Both Eqs. (26) and (39) can be parameterized and fitted to the general Eq. (3) as in Section 3. This means that a new self consistent cosmology has been developed because the same m function method has been used for both the solar system and the whirlpool galaxy. The binary pulsar can be modelled by the analytical function:

$$r = e^{\zeta\theta} \frac{\alpha}{1+\epsilon\cos(x\theta)} \tag{40}$$

which is a product of that of the spiral and precessing ellipse. In this case simple differentiation gives:

$$\frac{dr}{d\theta} = \zeta r + e^{\zeta\theta} \frac{x\alpha\sin(x\theta)}{\left(1+\epsilon\cos(x\theta)\right)^2} \tag{41}$$

and the binary pulsar is described simply by its characteristic m function. In the standard model the whirlpool galaxy led to the recession into mediaeval dark matter, and the binary pulsar was incorrectly described by the function (11) plus gravitational radiation arising from the wholly incorrect Einstein field equation.

Finally in this section a new approach to Newtonian dynamics is suggested as follows by starting with the infinitesimal line element:

$$ds^2 = c^2 d\tau^2 = m(r)c^2 dt^2 - \frac{dr^2}{m(r)} - r^2 d\theta^2. \tag{42}$$

In the solar system:

$$m(r) = \left(\frac{1}{b^2} - \left(\frac{x\epsilon}{\alpha}\right)^2 + \left(\frac{x}{\alpha}\right)^2 \left(1 - \frac{\alpha}{r}\right)^2\right)\left(\frac{1}{a^2} + \frac{1}{r^2}\right)^{-1} \tag{43}$$

which gives the precessing ellipse:

$$r = \frac{\alpha}{1+\epsilon\cos(x\theta)}. \tag{44}$$

When

$$x = 1 \tag{45}$$

Eq. (44) reduces to the static Newtonian ellipse:

$$r = \frac{\alpha}{1+\epsilon\cos\theta} \tag{46}$$

which from Euler Lagrange analysis is a solution of:

$$\frac{d^2u}{d\theta^2} + u = -\frac{m}{L^2u^2}F(u). \tag{47}$$

where

$$u = \frac{1}{r} \tag{48}$$

and F is the inverse square law of universal gravitation:

$$F = -\frac{mMG}{r^2} \tag{49}$$

incorrectly attributed to Isaac Newton. In historical fact, Robert Hooke suggested this law to Isaac Newton who developed it (John Aubrey, 'Brief Lives').

3. Computer analysis and orbital schematics

In this section the function $m(r)$ is evaluated. In order to give physical meaningful values, the limit

$$m(r \rightarrow \infty) = 1 \tag{50}$$

has to be fulfilled in all cases. This means that there is an additional relation between the parameters of m. We work this out for precessing orbits in the form of ellipses, logarithmic spirals and ellipses with shrinking diameter.

3.1 m functions for the solar system

The general form of $m(r)$ for precessing elliptic orbits was derived in Eq. (9) by equating the angular derivative of radius (Eq. (5)) with the general form of this expression (Eq. (6)) obtained from the infinitesimal line element. The limit of this function for $r \rightarrow \infty$ is

$$\lim_{r\rightarrow\infty} m(r) = -\frac{a^2(b^2\epsilon^2x^2 - b^2x^2 - \alpha^2)}{\alpha^2b^2}. \tag{51}$$

Setting this expression to 1, computer algebra gives the relation

$$\epsilon = \frac{\sqrt{a^2b^2x^2 - \alpha^2b^2 + a^2\alpha^2}}{abx} \tag{52}$$

(actually we used the positive solution of a square root). This expression can

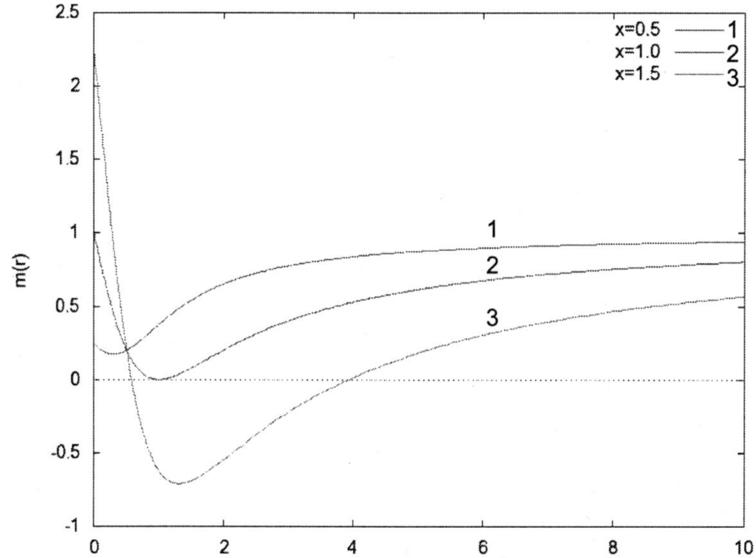

Fig. 1. $m(r)$ for precessing ellipses with parameters a = 1, α = 1.

be inserted in Eq. (9):

$$m(r) = -\frac{\left(2a^2r - a^2\alpha\right)x^2 - \alpha r^2}{\alpha r^2 + a^2\alpha}. \tag{53}$$

This is a simplified form of m where ϵ has been elimitated. Note that also the parameter b does not occur anymore, giving a dependence only from a, α and x. The curve is graphed in Fig. 1. The factor x determines the direction of precession of the ellipse. $x > 1$ means in directon of θ, $x < 1$ gives a rotation in the opposite direction. In the latter case, m remains in the positive range, meaning that there are no singularities in the metric.

Alternatively, the parameter x can be elimitated from Eq. (9) by using the limiting Eq. (51). This gives

$$x = \frac{\alpha\sqrt{b^2 - a^2}}{ab} \tag{54}$$

and the m function then takes the form

$$m(r) = \frac{\left(b^2\epsilon^2 - b^2\right)r^2 + \left(2\alpha b^2 - 2a^2\alpha\right)r - \alpha^2 b^2 + a^2\alpha^2}{\left(b^2\epsilon^2 - b^2\right)r^2 + a^2b^2\epsilon^2 - a^2b^2} \tag{55}$$

as an alternative to (53). The graph of this function is shown in Fig. 2 for three values of ϵ. It must be noted that this function looks different compared to Fig. 1, it rises slightly beyond unity and approaches unity from above in the r limit.

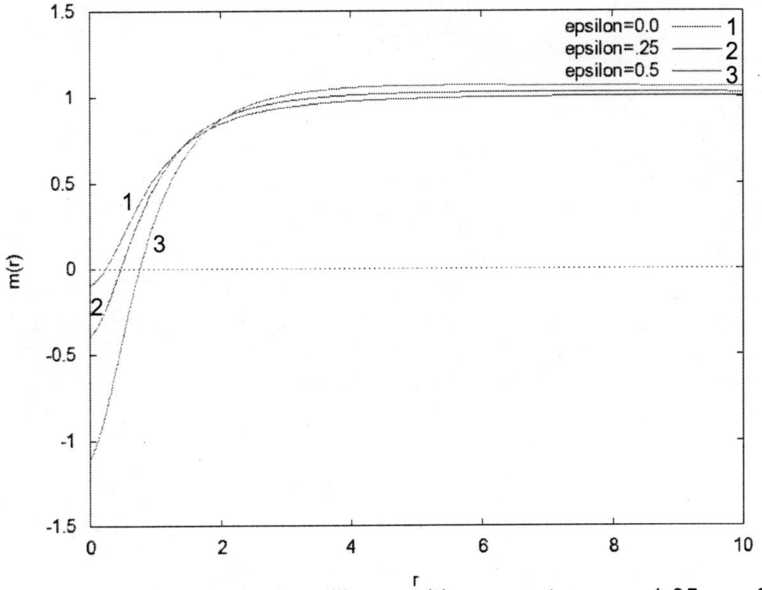

Fig. 2. $m(r)$ for precessing ellipses with parameters a = 1.05, α = 3.

3.2 m functions for whirlpool galaxies

The m function takes a simple form in case of logarithmic spirals. The calculation in section 2 leads to the m function of Eq. (39):

$$m(r) = \left(\frac{1}{b^2} - \frac{\zeta^2}{r^2}\right)\left(\frac{1}{a^2} + \frac{1}{r^2}\right)^{-1}. \tag{56}$$

This has the simple limit

$$\lim_{r\to\infty} m(r) = \frac{a^2}{b^2} \tag{57}$$

which means that

$$a \approx b \tag{58}$$

in the limit of large r, i.e. The kinetic energy plays no role compared to the relativistic total energy. The function (56) has been plotted in Fig. 3 for three 'pitch' values ζ. There is no difference if ζ is positive or negative because it appears in (39) in the squared form only. In the case ζ = 0 the orbit is a circle which has a purely positive m function.

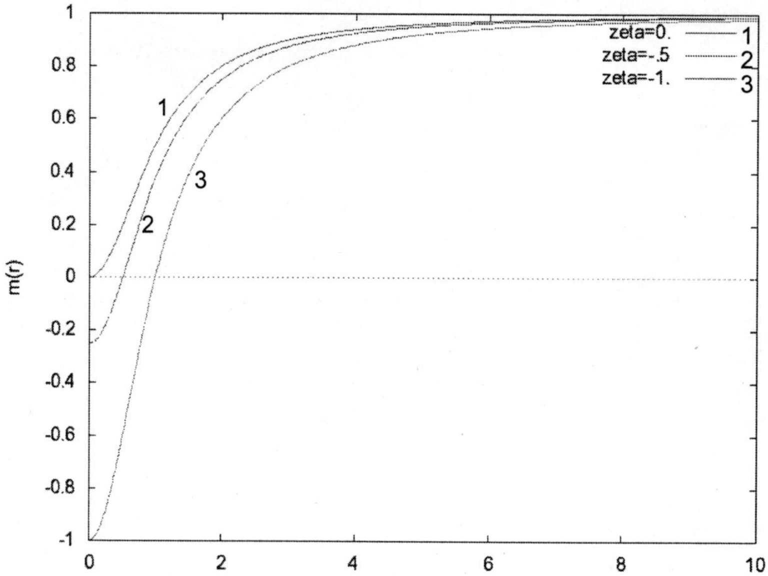

Fig. 3. $m(r)$ for logarithmic spirals with parameters $a = b = 1$.

3.3 m functions for binary pulsars

Massive cosmic objects in near distance to each other show a small permanent decrease of their average orbital radius. This was attributed to gravitational radiation losses prior to ECE theory. The orbits can be considered as a combination of an inward spiral with a precessing ellipse. The corresponding m function is given by Eq. (41). Applying the same limit calculation as before we obtain

$$\lim_{r\to\infty} m(r) = \frac{a^2 e^{-2\theta\zeta}\left(\alpha^2 e^{2\theta\zeta} - b^2\epsilon^2 x^2 + b^2 x^2\right)}{\alpha^2 b^2} \tag{59}$$

which is somewhat more complex than Eq. (51). Equating the limit to unity gives

$$\epsilon = \frac{\sqrt{-\alpha^2 b^2 e^{2\theta\zeta} + a^2\alpha^2 e^{2\theta\zeta}}}{abx} \tag{60}$$

and inserting this in (41) leads to an even more complicated expression:

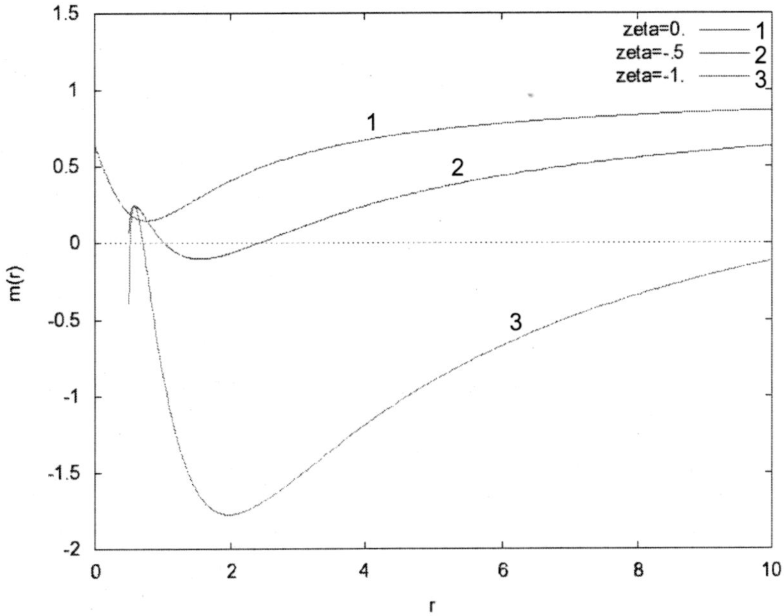

Fig. 4. $m(r)$ for shrinking precessing ellipses with parameters $a = 1.01$, $b = 1$, $\alpha = 1$, $\zeta = -1$, $\theta = \pi/4$.

$$
\begin{aligned}
m(r) = {} & -\frac{1}{\alpha b\left(r^2+a^2\right)} e^{-2\theta\zeta} \\
& \left(2ar\zeta e^{\theta\zeta}\sqrt{\left(a^2\alpha^2-\alpha^2 b^2\right)e^{2\theta\zeta}+a^2b^2x^2}\right. \\
& \sqrt{\left(\alpha^2b^2-a^2\alpha^2\right)r^2e^{2\theta\zeta}+\left(a^2\alpha^2b^2-2a^2\alpha b^2 r\right)x^2} \\
& \left.+\left(a^2\alpha b\zeta^2-\alpha b r^2\right)e^{2\theta\zeta}+\left(2a^2br-a^2\alpha b\right)x^2\right).
\end{aligned}
\tag{61}
$$

This expression depends on θ and has been graphed in Fig. 4 for a representative value of θ. The effect of the pitch is a significant drop of the m function to negative values. The limit for large r is unity again.

Alternatively we can solve the limiting equation (59) for x as before:

$$
x = \frac{\alpha\sqrt{b^2-a^2}\,e^{\theta\zeta}}{ab}. \tag{62}
$$

Inserting this into (41) gives a highly complicated expression again:

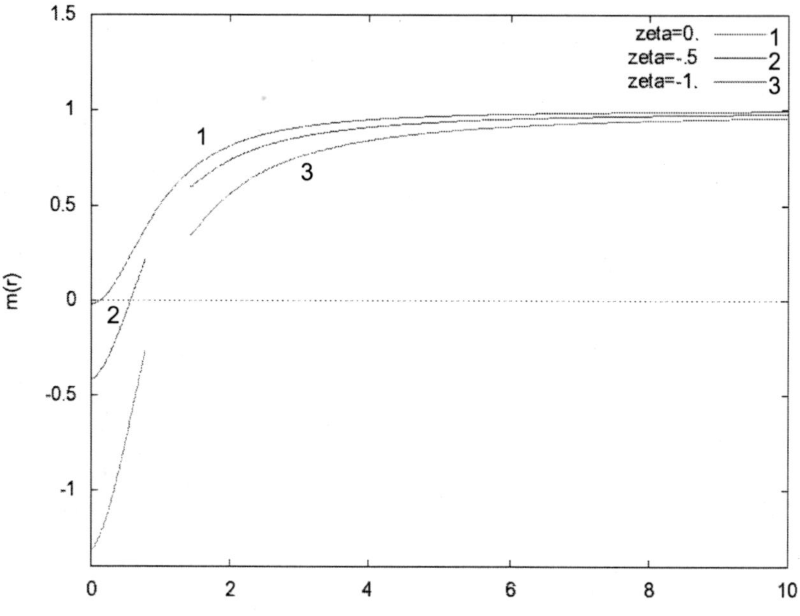

Fig. 5. $m(r)$ for shrinking precessing ellipses with parameters a = 1.01, b = 1, α = 1, ϵ = 0.3 .

$$
\begin{aligned}
m(r) = &-\frac{1}{\left(b^2\epsilon^2 - b^2\right)r^2 + a^2b^2\epsilon^2 - a^2b^2}\left(\left(a^2b^2\epsilon^2 - a^2b^2\right)\zeta^2\right. \\
&+\left(2ab\epsilon^2 - 2ab\right)\sqrt{-\frac{b^2 - a^2}{\epsilon^2 - 1}}\sqrt{\left(\epsilon^2 - 1\right)r^2 + 2\alpha r - \alpha^2\zeta} \\
&\left.+\left(b^2 - b^2\epsilon^2\right)r^2 + \left(2a^2\alpha - 2\alpha b^2\right)r + \alpha^2 b^2 - a^2\alpha^2\right).
\end{aligned}
\tag{63}
$$

This expression has the benefit of not to dependent on θ. However, the values are complex in a certain range of r. In that range no real values of m exist, as can be seen from Fig. 5. Occurence of complex values can easily be seen from Eq. (62) where the condition $a > b$ leads to a negative argument of the square root. Nevertheless the m functions looks more regular in overall than in Fig. 4. The same result can be drawn for the precessing ellipse of the solar system, see section 3.1, leading to a certain consistency of the results.

3.4 Graphical demonstration of elliptic orbits

Fig. 6: The Static Newtonian Elliptical Orbit describes an ellipse that, although non deviant, has a spiral connection in that all quadrants of the ellipse trace are almost exactly logarithmically spiral.

Fig. 7: The Precessing Elliptical Orbit trace can appear to produce spiralling "arms" from its rotating perihelion in both directions. Its rotation has slightly opened or closed the Newtonian, 360 degree, elliptical symmetry and so is now in spiralling elliptical orbit. The rotation of the orbit must depict a reduction

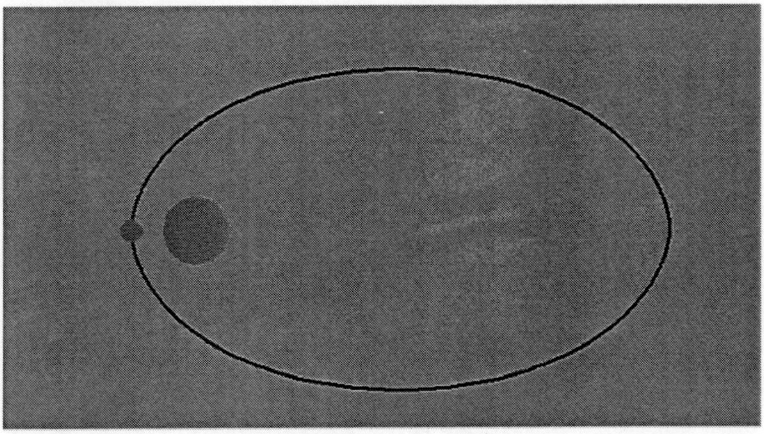

Fig. 6. Newtonian elliptical orbit.

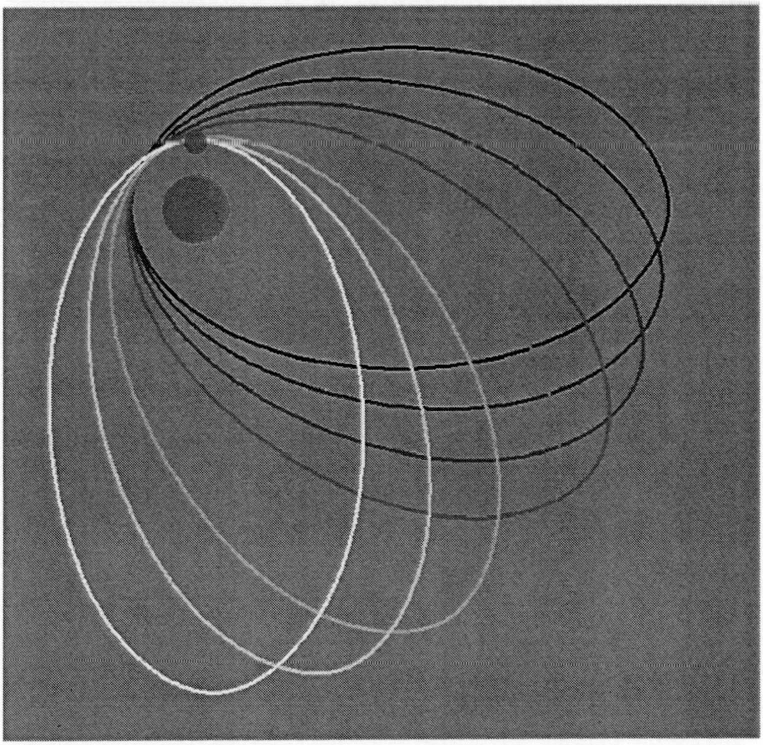

Fig. 7. Precessing elliptical orbit.

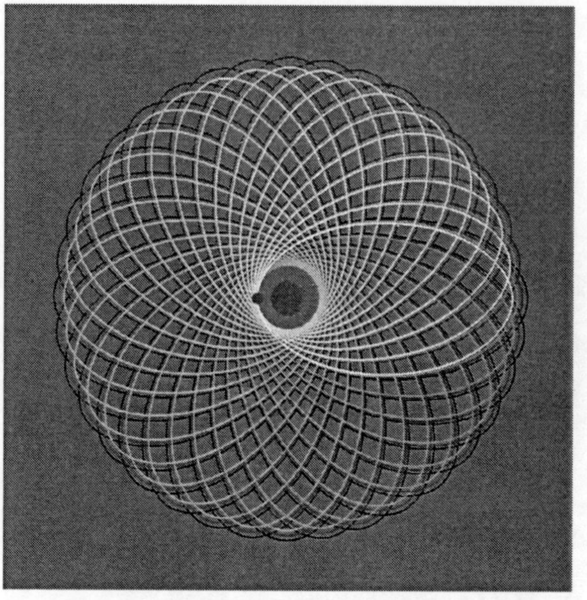

Fig. 8. Precessing and shrinking elliptical orbit.

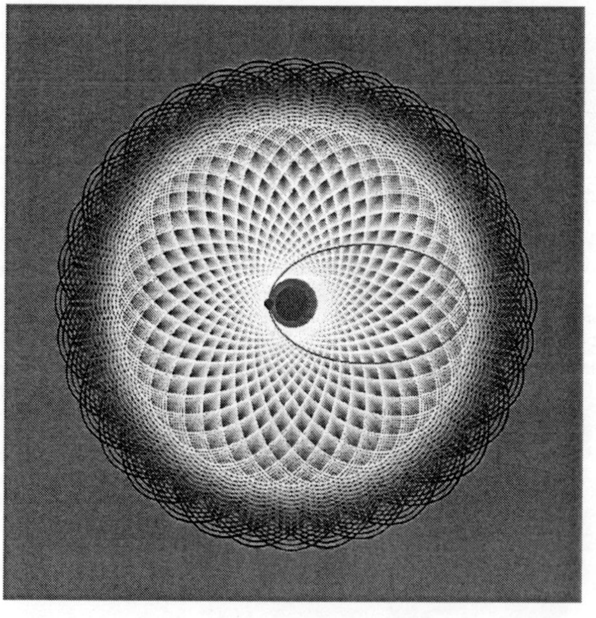

Fig. 9. Precessing and shrinking elliptical orbit.

or increase in the 360 degrees of the Newtonian Static ellipse's symmetry. So, more than 360 degrees of elliptical orbit gives a clockwise precession or vice versa if less than that (assuming a clockwise elliptical orbit).

Fig. 8: The Shrinking Precessing Elliptical Orbit trace is now seen to also spiral inward. There are many periods of orbit as there are precession rates. Therefore any one of these Figures shown may take up to millions of years to complete in reality.

Fig. 9: The Shrinking orbit continues toward M, its focus, possibly increasing in velocity, leading eventually to collision, diversion, or other uncertainty. A spiralling Galaxy can be depicted as a group of expanding precessing elliptical orbits that merely have a differing - much faster graphical spiral development.

References

[1] M. W. Evans, Ed., J. Foundations of Physics and Chemistry, (Cambridge International Science Publishing, bi monthly from June 2011).

[2] M. W. Evans, S. Crothers, H. Eckardt and K. Pendergast, "Criticisms of the Einstein Field Equation" (Cambridge International Science Publishing, Spring 2011).

[3] Kerry Pendergast, "The Life of Myron Evans" (Cambridge International Science Publishing, Spring 2011, available from www.cisp-publishing.com, Amazon, Ingram=s Catalog, W. H. Smith, The Guardian and all good bookshops).

[4] M .W. Evans, H. Eckardt and D. W. Lindstrom, AGenerally Covariant Unified Field Theory (Abramis, 2005 - 2011), in seven volumes.

[5] M .W. Evans, H. Eckardt and D. W. Lindstrom, plenary paper published by the Serbian Academy of Sciences, 2010; ibid. M. W. Evans and H. Eckardt, 2011.

[6] The ECE websites: www.aias.us (archived on www.webarchive.org.uk), www.atomicprecision.com, www.et3m.net, www.upitec.org.

[7] L. Felker, "The Evans Equations of Unified Field Theory" (Abramis, 2007), Spanish translation by Alex Hill on www.aias.us.

[8] M. W. Evans, Ed., "Modern Non-Linear Optics" (Wiley 2001, second edition), in three volumes; M .W. Evans and S. Kielich (eds.), "Modern Non-Linear Optics" (Wiley 1992, 1993, 1997, first edition) in three volumes.

[9] M .W. Evans and L. B. Crowell, "Classical and Quantum Electrodynamics and the B(3) Field" (World Scientific 2001).

[10] M. W. Evans and J.- P. Vigier, "The Enigmatic Photon" (Kluwer, 1994 to 2002) in ten volumes.

[11] A. A. Vankov, www.wbabin.net/eeuro/vankov.pdf.

[12] Miles Mathis, http://milesmathis.com/merc/html

Failure of the fundamentals of Einsteinian General Relativity

M.W. Evans[1*] and H. Eckardt[2]**

**Alpha Institute for Advanced Studies (AIAS) (www.aias.us); **Unified Physics Institute of Technology (UPITEC) (www.upitec.org)*

The incorrectness is demonstrated simply and straightforwardly of the basic claim of Einsteinian general relativity, that the so called "Schwarzschild" metric produces a precessing ellipse. The simplest type of lagrangian dynamics is sufficient to show that this claim cannot be true. It is argued that there is no general force law in cosmology, each system has its own lagrangian force law. Light bending and time delay due to gravitation are developed in terms of simple lagrangian dynamics, and this method is preferred to general relativity by Ockham's Razor.

Keywords: ECE theory, lagrangian dynamics, basic incorrectness of Einsteinian general relativity.

1. Introduction

In this series of papers developing ECE theory [1–10] it has been shown that Einsteinian general relativity is meaningless and riddled with fundamental errors. The principles of Einstein Cartan Evans (ECE) unified field theory have been used to begin the task of suggesting a new cosmology based on the antisymmetric connection. Einsteinian general relativity (EGR) has been heavily criticised for almost as long as it has existed, and is thoroughly obsolete. Not only is the subject riddled with errors, but also with poor historical scholarship. The most glaring mis attribution is the so called 'Schwarzschild metric', which is a fabrication, and is not due to Schwarzschild. In historical fact [11,12] the latter wrote to Einstein on Dec. 22nd 1915 to criticise the latter's claim to have calculated the precession of the perihelion. Schwarzschild pointed out that the calculation was incorrect and genuine scholarship has found that Schwarzschild was correct [11]. There are doubts [11, 12] concerning Einstein's personal integrity as a scientist, because he was under pressure to produce a result from general relativity. At that time very few accepted the theory.

[1]e-mail: EMyrone@aol.com
[2]e-mail: horsteck@aol.com

Later the Einstein field equation was criticised by Bauer, Schroedinger, Eddington, Levi-Civita, Dirac, Vigier and many others. In UFT 122 of this series (www.aias.us) it has been shown that the field equation is meaningless because of its use of an incorrect Christoffel connection. It is very easy to show that the connection is antisymmetric as in UFT 137, and not symmetric as in the Einstein field equation of 1915. In UFT 190 and 192 of this series, EGR was further criticised in other ways and new types of cosmology suggested. In Section 2 of this paper, the well known [13, 14] Euler Lagrange equations of any orbit in a plane are used to show that the force law of a precessing ellipse is the sum of an inverse square and inverse cube in the radial coordinate r. This result can be arrived at using the most elementary methods of lagrangian dynamics. The incorrect claim of EGR is that the precessing ellipse is due to the sum of inverse square and inverse fourth terms in r. Exactly the same lagrangian methods are used in EGR [13], which is therefore irretrievably incorrect and must be discarded. The claims of EGR to have described light deflection and time delay due to gravitation were heavily criticised in UFT 150 and 155, and these claims are complete nonsense. The lagrangian method is used in Section 2 of this paper to produce simpler theories of light bending and time delay due to gravitation. Furthermore, it is shown in Section 2 that there cannot be a universal law of gravitation as proposed by Hooke and communicated to Newton in the seventeenth century. The Newtonian gravitational law produces a static elliptical orbit, but the observed orbit in the solar system is a precessing ellipse. It is known now that the precessing ellipse is not due to EGR, which is well known to fail completely in an object such as a whirlpool galaxy. Dark matter has never been observed, and is a gross fallacy, big bang never occurred, and black holes do not exist. By Ockham's Razor therefore the description of cosmology by simple lagrangian dynamics is preferred. The lagrangian method is straightforward, it begins with the analytical function of the orbit, obtained directly from observation. Each orbit has its force law obtained from the two Euler Lagrange equations of any planar orbit [13,14]. There is no universal force law that describes all orbits.

In Section 3 the lagrangian cosmology is illustrated with examples such as the binary pulsar, and it is also demonstrated numerically that the force law of EGR cannot produce a stable precessing ellipse, it produces an unstable result that either shrinks or expands the orbit, and also modulates the orbit with other motions that are never observed. The analytical methods used in EGR [13,14] are therefore incorrect, they depend on false approximations. The actual analytical result of the EGR force law cannot be found, but in a rough approximation [13], it is a precessing ellipse modulated by a motion that is not in fact observed. This obviously erroneous result has been discarded for a century in a wholly unscientific manner. Therefore it is impossible for Einstein to have described the precession of the planet Mercury on 22nd Nov. 1915 [11, 12]. According to scholarship [11] this was his only paper on the subject, and caused a century of pseudoscience and endless confusion. Using the elegant eighteenth century methods of Lagrange brings us back on course to science.

2. Lagrangian force law of a precessing ellipse, light bending nd time delay by Lagrangian methods

From the two Euler–Lagrange equations of any planar orbit [13] the following well known equation can be deduced:

$$\frac{d^2}{d\theta^2}\left(\frac{1}{r}\right)+\frac{1}{r}=\frac{-mr^2}{L}F(r) \tag{1}$$

where m is the mass of the orbiting object, r is the radial distance between m and the attracting object M, L is the total angular momentum, a constant of motion [13], and $F(r)$ is the force of gravitational attraction between m and M. The force is:

$$F=\partial V/\partial r \tag{2}$$

where V is the gravitational potential. In Eq. (1) the cylindrical polar coordinates (r,θ) are used in a plane defined by:

$$dZ = 0. \tag{3}$$

Consider now the orbit of an object of mass m in the solar system, such as a planet, comet, or photon of mass m. Since ancient times this orbit has been observed by astronomers, and is a precessing ellipse. The equation of the precessing ellipse is:

$$\frac{\alpha}{r}=1+\epsilon\cos(x\theta) \tag{4}$$

where α is the half right latitude, ϵ is the eccentricity, and x is the precession constant. From Eq. (4) it follows that:

$$\frac{d^2}{d\theta^2}\left(\frac{1}{r}\right)=-\frac{x^2\epsilon}{\alpha}\cos(x\theta) \tag{5}$$

and that:

$$\frac{d^2}{d\theta^2}\left(\frac{1}{r}\right)+\frac{1}{r}=\frac{1}{\alpha}\left(1+\epsilon\left(1-x^2\right)\cos(x\theta)\right) \tag{6}$$

so the force of gravitational attraction is:

$$F(r) = -\frac{L^2}{m\alpha r^2}\left(1+\epsilon\left(1-x^2\right)\cos(x\theta)\right). \tag{7}$$

Using Eq. (4), the cosine term is:

$$\cos(x\theta) = \frac{1}{\epsilon}\left(\frac{\alpha}{r}-1\right) \tag{8}$$

so the force law is:

$$F(r) = \frac{-L^2}{mr^2}\left(\frac{x^2}{\alpha}+\frac{1}{r}\left(1-x^2\right)\right) \tag{9}$$

a result that was checked by computer.

This is the sum of an inverse square force law and an inverse cube force law. This is obviously the force law that produces stable precessing elliptical orbit as observed, and no other force law does so. The well known force law of attraction of EGR [1–10,12,14] is:

$$F(r) = ?\frac{-mMG}{r^2} - \frac{3MGL^2}{mc^2r^4}. \tag{10}$$

where G is Newton's constant, and this is incorrect. It does not produce a precessing ellipse because it is the sum of an inverse square and inverse fourth terms in r. This is a simple and conclusive demonstration that EGR is incorrect and meaningless.

When

$$x = 1 \tag{11}$$

the correct lagrangian force law (9) reduces to:

$$F(r) = -\left(\frac{L^2x^2}{m\alpha}\right)\frac{1}{r^2} \tag{12}$$

which is an inverse square law. The law (12) reduces to the inverse square law of Newton if:

$$\alpha = \frac{L^2}{m^2 MG}. \tag{13}$$

In historical fact, the law (12) was discovered by Robert Hooke and communicated to Isaac Newton, who developed Hooke's discovery (see online version of John Aubrey, 'Brief Lives', entry on Robert Hooke). For a circular orbit:

$$\alpha = r \tag{14}$$

and the force law for a circular orbit is an inverse cube law, not an inverse square law. There is no Newtonian law of universal gravitation. The lagrangian method developed in this paper is far more generally applicable.

For example, a binary pulsar system was considered in UFT 108 (www.aias.us) and consists of two objects of nearly equal mass orbiting each other in a precessing ellipse that is not stable. For a binary pulsar the orbit is shrinking. For purposes of illustration consider the orbit of a precessing ellipse that is spiralling outwards:

$$r = \frac{\alpha}{1+\epsilon\cos(x\theta)} e^{a\theta} \tag{15}$$

where a is the pitch of the spiral. From Eqs. (1) and (15) the lagrangian force law of attraction between the two masses of the binary pulsar is:

$$F(r) = -\frac{L^2}{mr^3}\left(\frac{1+a^2+\epsilon\left(1-x^2+a^2\right)\cos(x\theta)+2a\epsilon\sin(x\theta)}{1+\epsilon\cos(x\theta)}\right). \tag{16}$$

This force law is illustrated numerically in Section 3 and is a periodically modulated force law. Another example illustrated in Section 3 is that of a whirlpool galaxy, where the stars are arranged on a spiral. If this is a hyperbolic spiral, the force law is also a spiral, the mirror image of the star spiral. This is discussed further in Section 3, and this force law is again non Einsteinian and non Newtonian. As has been argued, even the simple circular orbit is non Newtonian.

If the mass m is a photon orbiting the sun, its orbit is:

$$\frac{d\theta}{dr} = \frac{\alpha}{xr}\left(\epsilon^2 r^2 - (\alpha - r)^2\right)^{-1/2} \tag{17}$$

an equation obtained by simple differentiation of the orbital equation (4).

$$r = \frac{\alpha}{1+\epsilon\cos(x\theta)}, \tag{18}$$

and is non-Einsteinian and non-Newtonian. The light deflection is obtained following the methods of UFT 150 (www.aias.us) and is

$$\Delta\theta = \frac{2\alpha}{x}\int_{R_0}^{\infty}\frac{1}{r}\left(\epsilon^2 r^2 - (\sigma - r)^2\right)^{-1/2} dr - \pi \tag{19}$$

where R_0 is the distance of closest approach. Using the transformation:

$$u = \frac{1}{r} \tag{20}$$

Eq. (19) simplifies to:

$$\Delta\theta = \frac{2\alpha}{x}\int_{\frac{1}{R_0}}^{0}\left(\epsilon^2 - (\alpha u - 1)^2\right)^{-1/2} du - \pi \tag{21}$$

In Section 3 Eq. (21) is integrated numerically to give a simple analytical result. Lagrangian dynamics can therefore describe light deflection due to gravitation, and in a correct, scientific way. The EGR claim is famous (or infamous) and wholly meaningless.

Similarly, (see accompanying note 193(7)) the time delay due to gravitation is given in lagrangian dynamics by:

$$\Delta t = t_3 - t_0 \tag{22}$$

where:

$$t_3 = \frac{2}{c}\left(\int_{R_0}^{R_E} f(r)\,dr + \int_{R_0}^{R_\rho} f(r)\,dr\right). \tag{23}$$

Here the time t_3 is given by:

$$f(r) = \left(\frac{m\alpha}{Lx}\right) r\left(\epsilon^2 r^2 - (\alpha - r)^2\right)^{-\frac{1}{2}} \tag{24}$$

and the time t_o by:

$$t_0 = 2(t_1 + t_2) \tag{25}$$

where

$$t_1 = \frac{1}{c}\int_{R_0}^{R_E}\left(1-\frac{R_0^2}{r^2}\right)^{-1/2} dr = \frac{r_1}{c}, \tag{26}$$

$$t_2 = \frac{1}{c}\int_{R_0}^{R_P}\left(1-\frac{R_0^2}{r^2}\right)^{-1/2} dr = \frac{r_2}{c}. \tag{27}$$

The Einsteinian result (see UFT 155 on www.aias.us) is again complete nonsense because its force law is incorrect, and for many other reasons. The lagrangian method is correct. The self consistency of time delay and light bending experiments should be tested carefully. The EGR results (UFT 155) showed no consistency at all.

3. Numerical analysis based on the Lagrangian method

The force laws and the result for light deflection derived in section 2 are studied in grater detail in this section.

3.1. Force law for binary pulsars

The force law (16) of binary pulsars depends on coordinates r and θ. These are not independent because the force relates to an orbit of the mass m. The orbital function $r(\theta)$ (Eq. (15)) has to be inserted into Eq. (16) to obtain a pure $F(r)$ dependence. The result obtained by computer algebra is

$$\begin{aligned} F(r) = & -\frac{e^{-a\theta}x^2L^2}{\alpha m r^2} \\ & -\frac{2ae^{-a\theta}\sqrt{(\epsilon+1)r-\alpha e^{a\theta}}\sqrt{(\epsilon-1)r+\alpha e^{a\theta}}\,xL^2}{} \\ & +\frac{(x^2-a^2-1)L^2}{mr^3} \end{aligned} \tag{28}$$

It is seen that there remains an exponential θ depemdence indicating some spiralling behaviour. The first term is essentially the Newtonian multiplied by an exponential. The result is plotted in Fig. 1. Parameters are chosen as indicated in the Figure caption. Three negative values were used for the pitch a of the decreasing elliptical orbit. The decrease of radius requires growing force over time or the angle θ respectively. This growth is exponential but modulated by the ellipse as can be seen from the figure. The orbit together with the force is shown in Fig. 2 in a polar diagram for $a = -0.01$. The decrease of the orbit and the rise of the force is manifest. The mass falls into the centre.

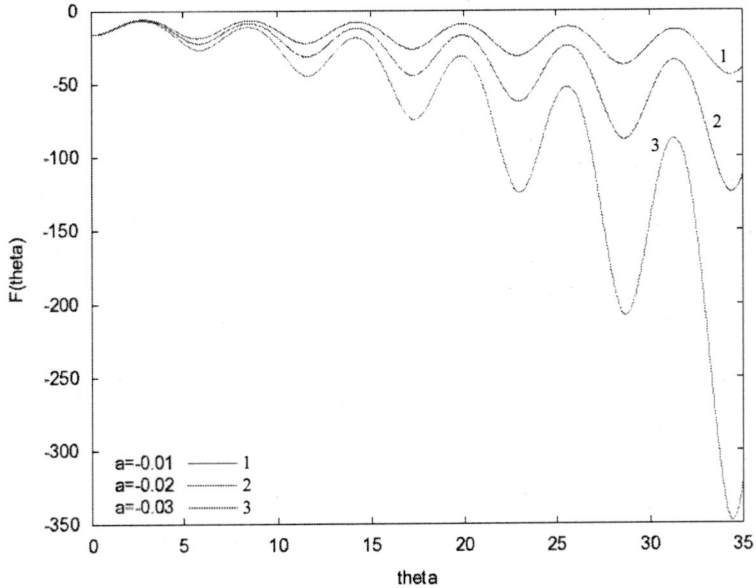

Fig. 1. Force for shrinking ellipses with parameters $L = 1$, $m = 0.1$, $- 1$, $x - 1.1$, $\epsilon - 0.3$.

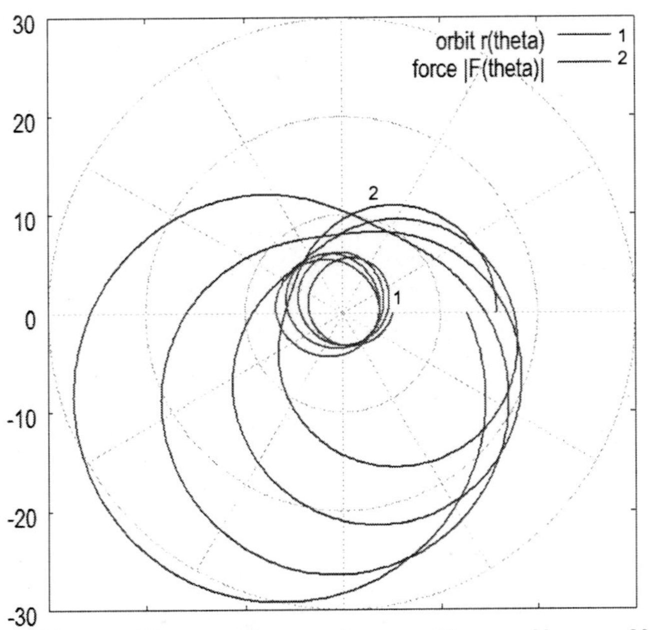

Fig. 2. Orbit and force for shrinking ellipses with parameters $L = 1$, $m = 0.1$, $\alpha = 1$, $x = 1.1$, $\epsilon = 0.3$, $a = -0.01$.

In a second example we used a positive parameter \$a=0.1\$ to describe the inverse case, an increasing orbit and falling force. This leads to a roughly logarithmic spiral in force as well as in orbit, see Fig. 3. Observe that the orbit pertains from inner to outer while the corresponding force runs from outer to inner.

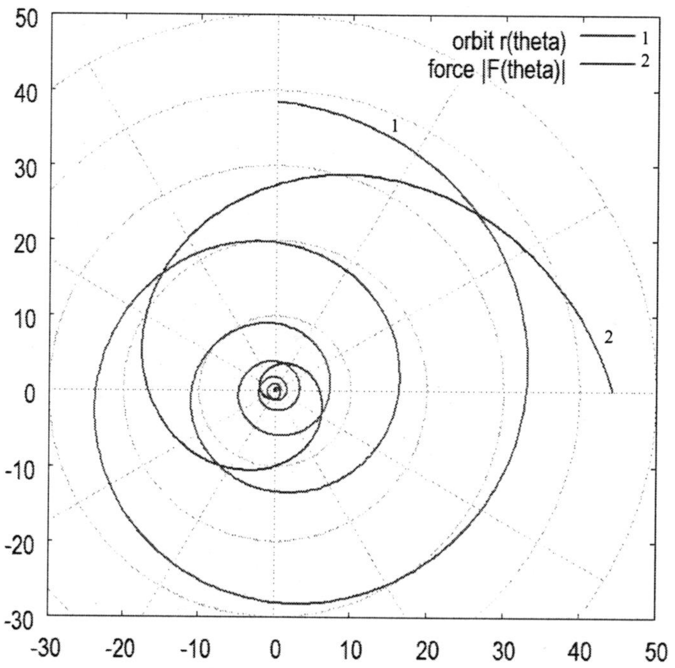

Fig. 3. Orbit and force for strongly expanding ellipses (spirals) with parameters L = 1, m = 0.1, α = 1, x = 1.1, ϵ = 0.3, a = 0.1.

3.2 Force law for whirlpool galaxies

The orbit of Fig. 3 may resemble a spiral galaxy, but we know that galaxies are formed more in a hyperbolic than a logarithmic spiral. The orbit has the simple form

$$r=\frac{\alpha}{\theta} \tag{29}$$

where α is a characteristic length and the spiral describes the linear asymptote for $\theta \to 0$. The direction of α is inverse to the direction of star movement from inner to outer. The Euler Lagrange equation (1) of this orbit is extremely simple since the second derivative of $1/r$ according to θ vanishes. The force law is

$$F=-\frac{\theta^3 L^2}{\alpha^3 m}, \tag{30}$$

i.e. F is proportional to θ^3, a third power force law. The graph (Fig. 4) shows the spiralling orbit (with the asymptote) and the force. Observe that for positive theta direction the orbit goes from outer to inner and the force from inner to outer. The polynomial spiral characteristic of the force is different to that of the orbit.

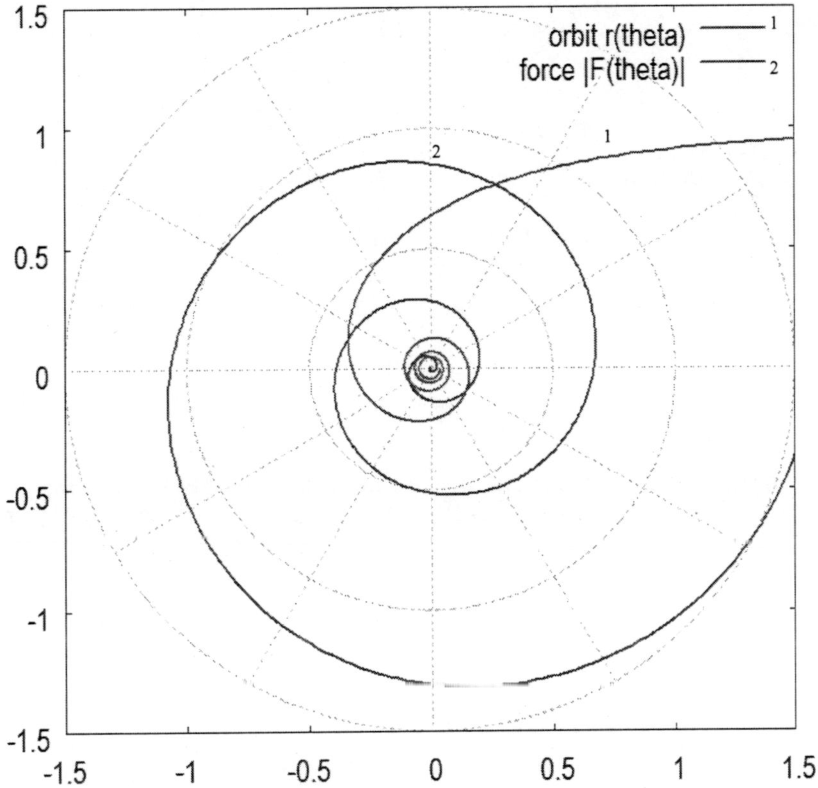

Fig. 4. Orbit and force for a hyperbolic spiral with parameters $L = 1$, $m = 0.1$, $\alpha = 1$.

3.3 Light deflection

The Lagrangian result for light deflection (21) can be handled analytically. The integral is in general

$$\int \frac{2\alpha}{x\sqrt{\epsilon^2-(\alpha u-1)^2}}du = -\frac{2}{x}\operatorname{asin}\left(\frac{2\alpha-2\alpha^2 u}{\sqrt{4\alpha^2(\epsilon^2-1)+4\alpha^2}}\right) = \frac{2}{x}\operatorname{asin}\left(\frac{\alpha u-1}{\epsilon}\right). \quad (31)$$

Therefore the angle of light deflection is

$$\Delta\theta = \frac{2}{x}\left(\operatorname{asin}\left(\frac{\alpha-R_0}{R_0\epsilon}\right)+\operatorname{asin}\left(\frac{1}{\epsilon}\right)\right). \quad (32)$$

For an orbit around the sun it is in good approximation $x \approx \epsilon \approx 1$. Then follows

$$\Delta\theta = -2\,\mathrm{asin}\left(\frac{\alpha - R_0}{R_0}\right) - 2\pi \tag{33}$$

or

$$\frac{\alpha}{R_0} - 1 = \sin\left(\frac{\Delta\theta}{2}\right) \tag{34}$$

from which follows

$$\alpha = R_0 \sin\left(\frac{\Delta\theta}{2}\right) + R_0. \tag{35}$$

Because is very small, the sinus term is nearly zero. This means that in good approximation

$$\alpha \approx R_0 \tag{36}$$

The characteristic radius α practically coincides with the sun radius.

Finally, we calculate the integrals of the time delay (23)–(27). Computer algebra yields

$$t_1 = \frac{\sqrt{R_E^2 - R_0^2}}{c}, \tag{37}$$

$$t_2 = \frac{\sqrt{R_P^2 - R_0^2}}{c}, \tag{38}$$

$$t_3 = \frac{2\sqrt{\alpha m}}{3cL}\Big(R_P\sqrt{2R_P - \alpha} + \alpha\sqrt{2R_P - \alpha} + R_E\sqrt{2R_E - \alpha} + \alpha\sqrt{2R_E - \alpha} - 2\sqrt{2R_0 - \alpha}\,\alpha - 2R_0\sqrt{2R_0 - \alpha}\Big). \tag{39}$$

The total time delay, formula (22),

$$\Delta t = t_3 - 2(t_1 + t_2), \tag{40}$$

contains more terms than the formula of Wald given by Eq. (29) of paper 155.

In total the Lagrangian method is suited to compute the time delay as well as all kind of gravitational forces.

References

[1] M. W. Evans, Ed., J. Found. Phys. Chem., (June 2011 onwards bimonthly), Cambridge International Science Publishing, www.cisp-publishing.com

[2] M. W. Evans, S. Crothers, H. Eckardt and K. Pendergast, "Criticisms of the Einstein Field Equation" (Cambridge International Science Publishing, 2011).

[3] K. Pendergast, "The Life of Myron Evans" (Cambridge International Science Publishing, 2011).

[4] M .W. Evans, H. Eckardt and D. Lindstrom, "Generally Covariant Unified Field Theory" (Abramis 2005 - 2011) in seven volumes.

[5] M. W. Evans, H. Eckardt and D. Lindstrom, "ECE Theory of H Bonding" (Serbian Academy of Sciences plenary, 2010), ibid. M. W. Evans and H. Eckardt.

[6] The ECE websites www.webarchive.org.uk, www.aias.us, www.atomicprecision.com, www.upitec.org, www.et3m.net..

[7] ECE papers in Found. Phys. Lett., Physica B, and Acta Physica Polonica.

[8] L. Felker, "The Evans Equations of Unified Field Theory" (Abramis 2007, Spanish translation by Alex Hill on www.aias.us).

[9] M. W. Evans and S. Kielich (Eds),"Modern Nonlinear Optics" (Wiley first and second editions, 1992, 1993, 1997, 2001, hardback, softback and e book), in six volumes; M. W. Evans and L. B. Crowell, "Classical and Quantum Electrodynamics and the B(3) Field" (World Scientific, 2001); M .W. Evans and A. A. Hasanein, "The Photomagneton in Quantum Field Theory" (World Scientific 1994).

[10] M. W. Evans and J.-P. Vigier, "The Enigmatic Photon" (Kluwer, 1994 to 2002, hardback and softback) in ten volumes.

[11] A. A. Vankov, www.wbabin.net/eeuro/vankov.pdf.

[12] Miles Mathis, http://milesmathis.com/merc/html.

[13] J. B. Marion and S. T. Thornton, "Classical Dynamics" (HBC, New York, 1988, third edition), chapter 7.

[14] S. P. Carroll, "Spacetime and Geometry: an Introduction to General Relativity" (Addison Wesley, New York, 2004).

Refutation of metric-based relativity for spherically symmetric spacetimes

M.W. Evans[1*] and H. Eckardt[2]**

**Alpha Institute for Advanced Studies (AIAS) (www.aias.us); **Unified Physics Institute of Technology (UPITEC) (www.upitec.org)*

It is shown straightforwardly that metric-based general relativity is meaningless in spherically symmetric spacetimes because it contains an irretrievable self inconsistency. The infinitesimal line element of a spherically symmetric spacetime is defined in terms of a function $m(r)$, but constancy of total energy and total angular momentum implies that $m(r)$ itself must be a constant independent of r. Therefore metric-based general relativity, a mainstay of twentieth century physics, must be abandoned. The only possibility now is that general relativity may be valid for the most general possible line element, but in that case the subject becomes untenable because of the complexity of the relation between metric and connection. Connection based general relativity as in ECE theory is still valid.

Keywords: refutation of metric-based general relativity, ECE theory.

1. Introduction

In recent papers of this series [1–10] it has been demonstrated in a simple and easily understandable way that Einsteinian general relativity cannot be a correct theory of physics. In the previous paper UFT 193 for example (see www.aias.us) it was shown that the Einsteinian general relativity (EGR) produces a force law that cannot describe a precessing elliptical orbit. The same methods as used in EGR were used in UFT 193 to produce the correct force law of attraction for a precessing elliptical orbit. This result means that Einstein could not possibly have produced the precession of Mercury as claimed endlessly in the dogmatic literature that has grown up around EGR. In UFT 150 and 155 it was shown that light bending by gravitation and gravitational time delay cannot be described by EGR. In Section 2 it is shown straightforwardly that metric-based general relativity in a spherically symmetric spacetime contains an irretrievable self inconsistency in that it is defined in terms of an r dependent function m(r), where r is the

[1]e-mail: EMyrone@aol.com
[2]e-mail: horsteck@aol.com

radial coordinate, but its own equation of motion requires the total energy E and total angular momentum L to be constants of motion. It is shown for the first time in this paper that the constancy of E and L implies the constancy of $m(r)$, a self contradiction at the most fundamental level. EGR is one of those theories that can be constructed in a spherically symmetric spacetime, and is invalidated completely and irretrievably by the constancy of $m(r)$.

In Section 3 the future directions and options remaining for general relativity are discussed. At present, metric-based general relativity could only be valid in the most general spacetime, where spherical symmetry is not assumed. However that would make the subject untenable due to the resulting complexity of the relation between the connection and metric. Connection based general relativity such as used in the field equations of ECE theory, the only available unified field theory, is still valid, because the metric is used only indirectly in raising and lowering indices. The standard model of physics and cosmology is refuted completely and irretrievably by this and other papers of the ECE series.

2. Self inconsistency of metric-based general relativity in a spherically symmetric spacetime

Consider the infinitesimal line element of a spherically symmetric spacetime [11]:

$$ds^2 = c^2 d\tau^2 = m(r)c^2 dt^2 - \frac{dr^2}{m(r)} - r^2 d\theta^2 \tag{1}$$

in cylindrical coordinates (r, θ) in the plane:

$$dZ^2 = 0. \tag{2}$$

In general [11] m is a function of r and t but for simplicity of argument and without loss of generality it can be considered to be a function of r. Here τ is the proper time, t is the time in the laboratory frame of the observer and c is the vacuum speed of light that is assumed to be a universal constant in general relativity. The total energy E and total angular momentum L are constants of motion defined by:

$$E = m(r)mc^2 \frac{dt}{d\tau} \tag{3}$$

and

$$L = mr^2 \frac{d\theta}{d\tau} = mr^2 \frac{d\theta}{dt}\frac{dt}{d\tau}. \tag{4}$$

By definition [12] the line element is:

$$ds^2 = c^2 d\tau^2 = m(r)c^2 dt^2 - d\mathbf{r}\cdot d\mathbf{r} \tag{5}$$

where:

$$d\mathbf{r}\cdot d\mathbf{r} = v^2 dt^2 = \frac{dr^2}{m(r)} + r^2 d\theta^2. \tag{6}$$

So in general:

$$c^2 d\tau^2 = \left(m(r)c^2 - v^2\right)dt^2 \tag{7}$$

and:

$$\frac{dt}{d\tau} = \left(m(r) - \frac{v^2}{c^2}\right)^{1/2}. \tag{8}$$

Therefore, the constant angular momentum is defined as follows:

$$L = mr^2 \frac{d\theta}{dt}\left(m(r) - \frac{v^2}{c^2}\right)^{-1/2}. \tag{9}$$

From previous work in this series [1–10] the angular velocity is defined by:

$$\omega = \frac{d\theta}{dt} = \frac{cbm(r)}{r^2} \tag{10}$$

where the constant b is defined by:

$$b = c\frac{L}{E}. \tag{11}$$

Therefore:

$$L = mcbm(r)\left(m(r) - \frac{v^2}{c^2}\right)^{-1/2} \tag{12}$$

which is the following quadratic for $m(r)$:

$$(mcb)^2 m^2(r) - L^2 m(r) + \frac{L^2 v^2}{c^2} = 0 \tag{13}$$

Its solution, checked by computer, is:

$$m(r) = \frac{1}{2}\left(\frac{E}{mc^2}\right)^2 \left[1 \pm \left(1 - \frac{4v^2}{c^2}\left(\frac{mc^2}{E}\right)^2\right)^{1/2}\right] \tag{14}$$

Similarly, the total energy E is a constant of motion defined by:

$$E = m(r)mc^2 \left(m(r) - \frac{v^2}{c^2}\right)^{-1/2}. \tag{15}$$

Solving this equation leads to Eq. (14) again self consistently. Using Eq. (6) the velocity in Eq. (15) is defined by:

$$v^2 = \frac{1}{m(r)}\left(\frac{dr}{dt}\right)^2 + r^2\left(\frac{d\theta}{dt}\right)^2. \tag{16}$$

In metric-based general relativity for any spherical spacetime:

$$\frac{dr}{dt} = cbm(r)\left(\frac{1}{b^2} - m(r)\left(\frac{1}{a^2} + \frac{1}{r^2}\right)\right)^{1/2} \tag{17}$$

and the angular velocity in any spherical spacetime is:

$$\frac{d\theta}{dt} = \frac{cbm(r)}{r^2}, \quad a = \frac{L}{mc}, \quad b = \frac{cL}{E}. \tag{18}$$

Therefore:

$$v^2 = c^2 m(r)\left(1 - \left(\frac{mc^2}{E}\right)m(r)\right) \tag{19}$$

where the following result has been used:

$$\frac{b}{a}=\frac{mc^2}{E}. \tag{20}$$

Solving Eqs. (19) and (14) for $m(r)$ gives the result:

$$m(r)=\left(\frac{E}{mc^2}-1\right)\left(1-\left(\frac{mc^2}{E}\right)^2\right)^{-1}=\left(\frac{E}{mc^2}\right)^2\left(1+\frac{E}{mc^2}\right)^{-1} \tag{21}$$

revealing for the first time in the history of metric-based general relativity that it is fatally flawed by an irretrievable self contradiction, because $m(r)$ from Eq. (21) is a constant in any spherical spacetime. Originally, $m(r)$ has been assumed to be a function of r, Q.E.D. In EGR the $m(r)$ function is incorrectly said to be:

$$m(r)=?\,1-\frac{r_0}{r} \tag{22}$$

where:

$$r_0=\frac{2MG}{c^2}. \tag{23}$$

Here G is Newton's constant. As shown in UFT 102 ff., the $m(r)$ function needed to describe a precessing ellipse or a whirlpool galaxy is a complicated function of r, and not the constant demanded by the constancy of E and L. So the metric-based general relativity in a spherical spacetime must be abandoned as a theory of natural philosophy.

The theory of special relativity can be forced to emerge from Eq. (21) in the limit:

$$E\rightarrow\gamma mc^2 \tag{24}$$

where:

$$\gamma=\left(1-v^2/c^2\right)^{-1/2}. \tag{25}$$

In the limit:

$$\frac{v}{c}\ll 1 \tag{26}$$

the constant $m(r)$ function becomes the indeterminate ratio:

$$m(r) \rightarrow \frac{0}{0} = 1. \tag{27}$$

So this is not generally valid because it applies only when $v \ll c$. So, it has been shown that metric-based general relativity does not reduce correctly to special relativity.

Therefore the gravitational sector of the standard model has been refuted completely.

3. The future of general relativity

It may be possible to devise a metrical theory of general relativity in the most general spacetime without assuming spherical symmetry. The aim would be to produce observed orbits from the infinitesimal line element of the most general spacetime. Such a theory would be very obscure and complicated, notably in the relation between metric and connection, and it is not known whether or not it would reduce correctly to special relativity. In UFT 193 classical lagrangian methods were shown to give a precessing ellipse, light bending and gravitational time delay without the use of any kind of relativity and that is a method of classifying observable orbits in terms of force laws. There appears to be no universal force law for all cosmological objects and systems. Connection based general relativity originally produced the Einstein field equation via the so called 'second Bianchi identity' [11]. However recent work [1–10] has uncovered numerous irretrievable errors in those methods of circa 1915, notably the neglect of torsion. The correct geometry must be one which uses torsion, and an adequate geometry is the one due to Cartan upon which ECE theory is based directly. The connection used in Cartan's type of geometry is the spin connection, and its two fundamental equations are the Cartan–Maurer structure equations which define the torsion form and the curvature form with the Cartan tetrad and the spin connection. This geometry is valid in any mathematical space and any dimension.

It is known following the simple but conclusive argument in Section 2 of this paper that the space of ECE theory cannot be spherically symmetric. An attempt to construct a unified field theory based on relativity cannot be based on a spherical spacetime and any attempt must be based on the connection based field equations of ECE. These are structurally identical for electrodynamics and dynamics. So the aim of future work will include attempts at deriving orbital theory from the ECE field equations without the use of an infinitesimal line element.

References

[1] M. W. Evans, S. Crothers, H. Eckardt and K. Pendergast, "Criticisms of the Einstein Field Equation" (Cambridge International Science Publishing, www.cisp-publishing.com. , 2011).

[2] M .W. Evans, Ed., Journal of Foundations of Physics and Chemistry, from June 2011 bimonthly.

[3] Kerry Pendergast, "The Life of Myron Evans" (Cambridge International Science Publishing, Spring 2011), available from all good bookshops and online.

[4] M .W. Evans, H. Eckardt and D. W. Lindstrom, "Generally Covariant Unified Field Theory" (Abramis Academic, 2005 - 2011) in seven volumes.

[5] L. Felker, "The Evans Equations of Unified Field Theory" (Abramis Academic,2007, Spanish version by Alex Hill on www.aias.us).

[6] The ECE websites: www.webarchive.org.uk, www.aias.us, www.atomicprecision.com, www.upitec.org, www.et3m.net)

[7] M .W. Evans, H. Eckardt and D. W. Lindstrom,"ECE Theory Applied to H Bonding" (Plenary Paper, Serbian Academy of Sciences, 2010).

[8] M. W. Evans and S. Kielich, Eds., "Modern Nonlinear Optics" (Wiley, New York in six volumes and two editions, 1992, 1993, 1997, 2001), hardback, softback and e book.

[9] M .W. Evans and L. B. Crowell, "Classical and Quantum Electrodynamics and the B(3) Field" (World Scientific, 2001).

[10] M. W. Evans and J.-P. Vigier, "The Enigmatic Photon" (Kluwer, 1994 - 2002) in five volumes hardback, five volumes softback.

[11] S. - P. Carroll, "Spacetime and Geometry - an Introduction to General Relativity" (Addison Wesley, New York, 2004).

[12] J. D. Jackson, "Classical Electrodynamics" (Wiley 1999, 3rd Ed.)

A new cosmology with the Crothers metric

M.W. Evans[1*] and H. Eckardt[2]**

**Alpha Institute for Advanced Studies (AIAS) (www.aias.us); **Unified Physics Institute of Technology (UPITEC) (www.upitec.org)*

It is shown that the Crothers metric for a general spacetime is the only valid metric of line element based general relativity. The equation of motion, orbital equation, deflection of light due to gravitation, and various orbital properties of the metric are evaluated.
In general five experiments are needed to determine the unknowns of the Crothers metric, which however does not suffer from the severe internal inconsistencies of Einsteinian general relativity or relativity in a spherical spacetime.

Keywords: ECE theory, cosmology with the Crothers metric.

1. Introduction

Recently in this series of papers [1–10] it has been shown that the Einsteinian general relativity (EGR) and general relativity in a spherical spacetime suffer from a severe self inconsistency which invalidate them completely as theories of physics. These theories are based on the infinitesimal line element method. In Section 2 it is shown that the Crothers metric of general spacetime does not suffer from this defect, and is a valid theory of cosmology. The equation of motion, orbital equation, light deflection due to gravitation, and various orbital properties are of the Crothers metric are derived in this section. In Section 3 the results are checked by computer and a numerical analysis of the results suggested. In general the Crothers metric depends on five unknown parameters, so five independent experiments are needed to find them. Line element general relativity ceases to be a predictive theory of physics, and this work overturns a century of dogmatism in physics and cosmology.

2. Properties of the Crothers cosmology

In cylindrical polar coordinates in the plane:

[1]e-mail: EMyrone@aol.com
[2]e-mail: horsteck@aol.com

$$dZ = 0 \tag{1}$$

the infinitesimal line element of Crothers [1–10] is:

$$ds^2 = c^2 d\tau^2 = AC^{1/2} c^2 dt^2 - BC^{1/2} dr^2 - C(r) d\theta^2 \tag{2}$$

where A and B are unknown parameters and where C is defined by:

$$C(r) = \left(|r - r_0|^n + \alpha^n\right)^{2/n}; \tag{3}$$

here R is the radius of curvature and

$$r_0 = \frac{2MG}{c^2} \tag{4}$$

where G is Newton's constant, M is the attracting mass, and c the speed of light. The parameter α is another unknown and n is an integer. The radius of curvature R is related to the radial coordinate r. If m is the mass of the attracted object the Lagrangian from Eq. (2) is:

$$\mathcal{L} = \frac{1}{2} mc^2 = \frac{m}{2}\left(c^2 AC^{1/2}(r)\left(\frac{dt}{d\tau}\right)^2 - BC^{1/2}(r)\left(\frac{dr}{d\tau}\right)^2 - C(r)\left(\frac{d\theta}{d\tau}\right)^2 \right) \tag{5}$$

where τ is the proper time. The Euler Lagrange equation shows that the total energy is:

$$E = mc^2 AC^{1/2}(r)\frac{dt}{d\tau} \tag{6}$$

and the total angular momentum is:

$$L = mC(r)\frac{d\theta}{d\tau}. \tag{7}$$

Both E and L are constants of motion. By definition:

$$d\mathbf{r} \cdot d\mathbf{r} = v^2 dt^2 = BC^{1/2}(r) dr^2 + C(r) d\theta^2 \tag{8}$$

so the total velocity of m is defined by:

$$v^2 = BC^{1/2}(r)\left(\frac{dr}{dt}\right)^2 + C(r)\left(\frac{d\theta}{dt}\right)^2. \tag{9}$$

It follows that:

$$c^2 d\tau^2 = \left(c^2 AC^{1/2}(r) - v^2\right)dt^2, \tag{10}$$

$$\frac{dt}{d\tau} = \left(AC^{1/2}(r) - \frac{v^2}{c^2}\right)^{-1/2} \tag{11}$$

and that:

$$E = mc^2 AC^{1/2}(r)\left(AC^{1/2}(r) - \frac{v^2}{c^2}\right)^{-1/2}. \tag{12}$$

Therefore, Eq. (5) is:

$$mc^2 = \frac{E^2}{Amc^2 C^{1/2}(r)} - mBC^{1/2}(r)\left(\frac{dr}{d\tau}\right)^2 - \frac{L^2}{mC(r)}, \tag{13}$$

and this equation may be rearranged to give the equation of motion of the Crothers metric:

$$\frac{1}{2}m\left(\frac{dr}{d\tau}\right)^2 = \frac{1}{2BC(r)}\left(\frac{E^2}{Amc^2} - C^{1/2}(r)\left(mc^2 + \frac{L^2}{mC(r)}\right)\right). \tag{14}$$

The orbital equation is obtained using:

$$\frac{dr}{d\tau} = \frac{dr}{d\theta}\frac{d\theta}{d\tau}, \quad \frac{d\theta}{d\tau} = \frac{L}{mC(r)}, \tag{15}$$

from which it follows that:

$$\left(\frac{dr}{d\theta}\right)^2 = \frac{mC(r)}{BL^2}\left(\frac{E^2}{Amc^2} - C^{1/2}(r)\left(mc^2 + \frac{1}{C(r)}\frac{L^2}{m}\right)\right). \tag{16}$$

For a distance of closest approach R_0 the light deflection due to gravitation is:

$$\Delta\theta = 2\int_0^{R_0}\left(\frac{mC(r)}{BL^2}\right)^{1/2}\left(\frac{E^2}{Amc^2} - C^{1/2}(r)\left(mc^2 + \frac{1}{C(r)}\frac{L^2}{m}\right)\right)^{-1/2} dr \tag{17}$$

and depends on A, B, C, E and L. Following upon the collapse of EGR, the light deflection due to gravitation cannot be predicted by general relativity. In general five independent experiments are needed to determine the five unknowns A, B, C, E and L. One of these could be light deflection due to gravitation.

The orbital angular velocity of the object m around M is easy to measure and can be calculated by first calculating:

$$v = \frac{dr}{dt} = \frac{dr}{d\theta}\frac{d\theta}{d\tau}\frac{d\tau}{dt} \tag{18}$$

to give:

$$\frac{dr}{dt} = \frac{Lc^2A}{EC^{1/2}(r)}\frac{dr}{d\theta}. \tag{19}$$

So the orbital angular velocity is:

$$\omega = \frac{d\theta}{dt} = \frac{d\theta}{dr}\frac{dr}{dt} = \frac{Lc^2A}{EC^{1/2}(r)} \tag{20}$$

giving a second independent experiment. It is possible to measure experimentally in astronomy the dependence of θ on r, giving a third experiment. The fourth experiment consists of measuring the dependence of r on t, and the fifth experiment the measurement of the total orbital velocity v in Eq. (9). The measurement of the precession of the perihelion is not an unequivocal experiment in the solar system because the precession is very tiny and prone to uncertainties as is well known. It has become clear in recent work that Einstein did not predict this precession correctly, neither did he predict the deflection of light due to gravitation or the gravitational time delay.

Line element general relativity is no longer a predictive method of physics, and experiments to 'verify' Einstein's incorrect theory are meaningless. It might be more straightforward to develop a cosmology based on the field equations of ECE theory, using the spin connection rather than the metric and this will be the subject of future work.

3. Numerical analysis

As an example for the parameters of the Crothers metric we consider a precessing elliptical orbit. This is given by

$$r = \frac{\alpha}{\epsilon \cos(x\theta) + 1} \tag{21}$$

and the derivative of the orbit is

$$\frac{dr}{d\theta} = \frac{rx}{\alpha}\sqrt{\epsilon^2 r^2 - (\alpha - r)^2}. \tag{22}$$

Equating the square of this expression with Eq. (16), the general expression for the squared orbital derivative gives

$$\frac{r^2 x^2}{\alpha^2}\left(\epsilon^2 r^2 - (\alpha - r)^2\right) = \frac{mD(r)^2}{BL^2}\left(\frac{E^2}{c^2 mA} - D(r)\left(\frac{L^2}{D(r)^2} + c^2 m\right)\right) \tag{23}$$

where we have defined

$$D(r) := \sqrt{C(r)}. \tag{24}$$

Solving this equation for $D(r)$ gives an equation of third order. The three solutions are complex-valued and highly complicated but can be handled by computer algebra. The real parts of the solutions are graphed in Fig. 1. They are in part negative so that the function $C(r)$ is imaginary in those regions. We plotted the real values of $C(r)$ in Fig. 2. It can be seen that only the third solution is regular for the whole r range so that it can be assumed that this is the physically relevant solution.

The angular frequency given by Eq. (20) is proportional to the inverse square root of $C(r)$. This is graphed in Fig. 3. From this plot it is quite obvious that only the third solution is physical.

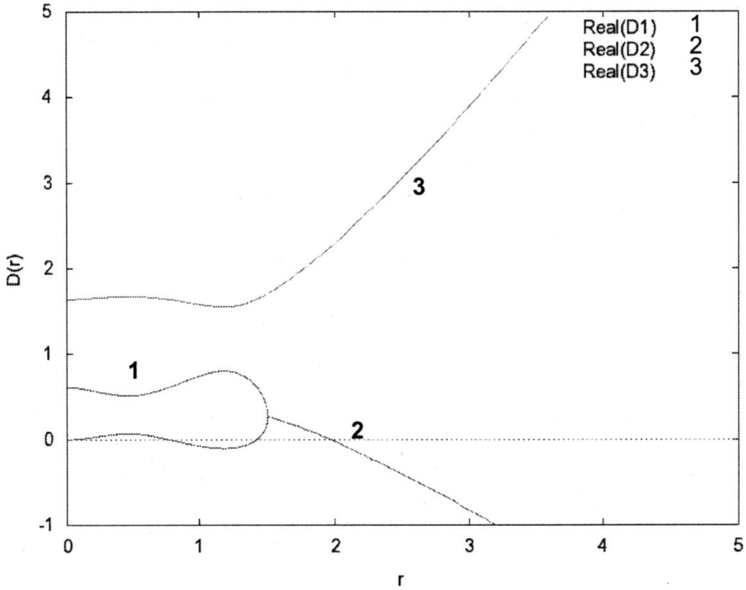

Fig. 1. Real part of the solutions for $D(r)$ for parameters $\alpha = x = A = B = m = c = L = 1$, $\epsilon = 0.3$, $E = 1.5$.

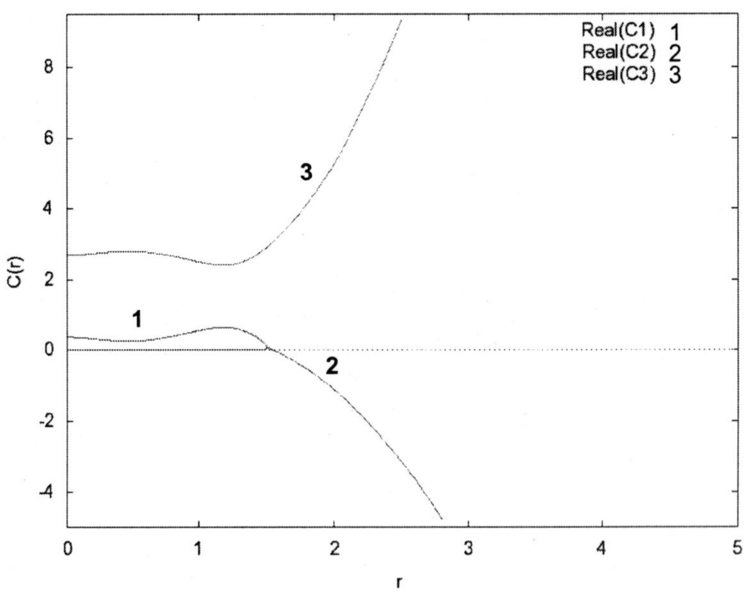

Fig. 2. Solutions $C(r)$.

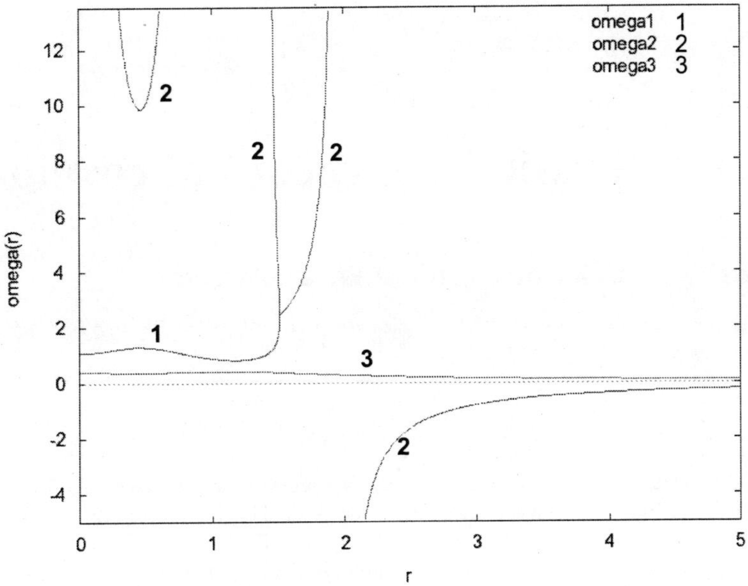

Fig. 3. Angular velocities corresponding to $C(r)$.

References

[1] M.W. Evans, S. Crothers, H. Eckardt and K. Pendergast, "Criticisms of the Einstein Field Equation" (Cambridge International Science Publishing, www.cisp-publishing.com , 2011).

[2] M. W. Evans, Ed., J. Found. Phys. Chem., bimonthly from June 2011.

[3] The ECE websites www.aias.us, www.atomicprecision.com, www.et3m.net, www.upitec.org.

[4] M .W. Evans, H. Eckardt and D. W. Lindstrom, "Generally Covariant Unified Field Theory" (Abramis, 2005 to 2011) in seven volumes.

[5] L. Felker, "The Evans Equations of Unified Field Theory" (Abramis 2007).

[6] K. Pendergast, "The Life of Myron Evans" (CISP 2011).

[7] M.W. Evans and S. Kielich (Eds.), "Modern Nonlinear Optics" (Wiley, 1992, 1993, 1997, 2001, first and second editions), in six volumes.

[8] M.W. Evans and L.B. Crowell, "Classical and Quantum Mechanics and the B(3) Field" (World Scientific, 2001).

[9] M.W. Evans and J.-P. Vigier, "The Enigmatic Photon" (Kluwer, Dordrecht 1994 to 2002, hardback and softback), in ten volumes.

[10] M.W. Evans, H. Eckardt and D.W. Lindstrom, several papers published in 2010 and 2011 by the Serbian Academy of Sciences.

A fully relativistic ECE theory of cosmology

M.W. Evans[1]*, H. Eckardt[2] and D.W. Lindstrom[3]***

**Alpha Institute for Advanced Studies (AIAS) (www.aias.us); **Unified Physics Institute of Technology (UPITEC) (www.upitec.org)*

A fully relativistic theory of cosmology is suggested on the concept of position tetrad. The linear velocity is defined in terms of the position tetrad and the spin connection. Using a simple antisymmetry law of ECE theory the expression for velocity is simplified into one containing one component of spin connection. The procedure of special relativity is used to define a new type of relativistic kinetic energy which is used in lagrangian theory to produce a force equation for any orbit.

Keywords: ECE theory, relativistic theory of cosmology, force law for any orbit.

1. Introduction

In recent papers of this series of one hundred and ninety five papers to date [1-10] the Einsteinian general relativity has been refuted directly in several different and very simple ways, and the refutations checked by computer for algebraic correctness. There can be no doubt that the era of twentieth century general relativity is over. Every one of its conclusions has been shown to be incorrect either by simple refutation as in UFT 194 (www.aias.us) or by astronomy. It has been known since December 1915 [11] that Einstein's theory of the perihelion precession has always been incorrect, and his theory of general relativity has been criticised repeatedly by some of the most eminent physicists and mathematicians, notably: Schwarzschild, Schrödinger, Eddington, Levi-Civita, Cartan, Dirac, Vigier and many others. The astronomical discovery of the whirlpool galaxy's velocity curve about half a century ago meant that the Einstein theory does not describe cosmology at all. It has persisted in the literature due to careless repetition of error, and has become dogmatism.

In Section 2 a new general relativity is initiated on the basis of Cartan geometry, and is based on torsion as is the whole of ECE theory [1–10]. The

[1]e-mail: EMyrone@aol.com
[2]e-mail: horsteck@aol.com
[3]e-mail: dwlindstrom@gmail.com

Cartan tetrad is used to define the position tetrad, from which the linear velocity is defined using the first structure equation of Cartan [12]. Simple use of an ECE antisymmetry law reduces the expression for linear velocity to one in one scalar component of the spin connection. This procedure is similar to one that uses the idea of covariant derivative, and is one that does not use any of Einstein's incorrect mathematics. The procedure of special relativity is extended to general relativity using a work integral to define the relativistic kinetic energy in terms of a characteristic evolution time t *f*, in terms of the spin connection and its radial derivative. Lagrangian dynamics are used to define the force law for any orbit.

In Section 3 the non-relativistic limit of this theory is evaluated using the cylindrical polar coordinates in a plane, and the same result found as in recent papers for the force law of a precessing ellipse. The force law is not that claimed in Einsteinian theory to produce a precessing ellipse. The Einstein theory uses the same lagrangian method. A direct check by computer algebra in UFT 192 and UFT 193 shows that the claimed force law of Einsteinian theory produces a very complicated orbit that is not a precessing ellipse at all. Therefore Schwarzschild's original and severe criticism of Einstein in December 1915 [11] was correct. Einstein's theory should have been abandoned at that time. Unfortunately Eddington claimed to have verified the theory and it persisted as unscientific dogma even after the discovery of the completely non Einsteinian velocity curve of a whirlpool galaxy no less than fifty years ago.

2. Towards a new relativity

In classical, non relativistic, dynamics the position of a particle is described by the position vector **r**. In Cartesian coordinates [13,14] its linear velocity (**v**) and linear acceleration (**a**) are described by:

$$\mathbf{v} = \frac{d\mathbf{r}}{dt}, \quad \mathbf{a} = \frac{d\mathbf{v}}{dt} \tag{1}$$

respectively, and the force **F** is defined for a particle of mass m as:

$$\mathbf{F} = m\mathbf{a}. \tag{2}$$

These are the familiar definitions of Newtonian dynamics. In the latter type of dynamics, the use of cylindrical polar coordinates means [13, 14] that the above definitions are changed to:

$$\mathbf{r} = r\mathbf{e}_r \tag{3}$$

$$\mathbf{v} = \dot{r}\mathbf{e}_r + r\dot{\theta}\mathbf{e}_\theta \tag{4}$$

$$\mathbf{a} = \left(\ddot{r} - r\dot{\theta}^2\right)\mathbf{e}_r + \left(r\ddot{\theta} + 2\dot{r}\dot{\theta}\right)\mathbf{e}_\theta \tag{5}$$

in the plane

$$dZ = 0. \tag{6}$$

The cylindrical polar coordinate system in this plane is described by (r, θ). The unit vectors of the cylindrical polar system [14] are:

$$\mathbf{e}_r = \mathbf{i}\cos\theta + \mathbf{j}\sin\theta \tag{7}$$

$$\mathbf{e}_\theta = -\mathbf{i}\sin\theta + \mathbf{j}\cos\theta. \tag{8}$$

In Cartesian coordinates in the plane (6):

$$\mathbf{r} = r_X\mathbf{i} + r_Y\mathbf{j} \tag{9}$$

and:

$$\mathbf{v} = \dot{\mathbf{r}} = \frac{d\mathbf{r}}{dt}, \quad \mathbf{a} = \dot{\mathbf{v}} = \frac{d\mathbf{v}}{dt}. \tag{10}$$

The reason why the Cartesian and cylindrical polar systems look so different [13] is that in the latter system the coordinate axes move. In the description of orbits the cylindrical polar system is used to give very simple equations for an ellipse:

$$r = \frac{\alpha}{1 + \epsilon\cos\theta} \tag{11}$$

and a precessing ellipse:

$$r = \frac{\alpha}{1 + \epsilon\cos(x\theta)}. \tag{12}$$

Here 2α is known as the right latitude (latus rectum), ϵ is the eccentricity, and x the precession constant. As shown in Section 3, the force law for an elliptical orbit and for a precessing elliptical orbit can be worked out using only Eqs. (1) to (12). It is shown in Section 3 that the result is the same exactly as in the classical lagrangian dynamics of UFT 193. In seventeenth century language, 'the force law of attraction' for a static ellipse is the Hooke Newton inverse square law, discovered by Robert Hooke [15] and not Isaac Newton as in the dogma. The force law of attraction for a precessing ellipse is a sum of inverse square and inverse cubed terms in r. In the incorrect dogma of Einstein theory [13] this sum is claimed to be that of an inverse square and inverse fourth power terms in r. In the Einstein method [13] an effective potential is defined and the same lagrangian method used as in UFT 193. Clearly, the Einstein dogmatists never bothered to check that their work was correct, their claimed sum of terms leads to a very complicated curve that is not a precessing ellipse at all (UFT 193 on www.aias.us). This is an astonishing illustration of how dogma can damage physics, and how useless a system of physics 'administration' can be. Theories can neither be administered nor proclaimed. Section 3 shows that even in the very familiar non relativistic context, the very concept of force law of attraction is untenable, and this can be shown in a very simple way simply by working out **v** and **a** with cylindrical polar coordinates in a plane. What has been known for three hundred and fifty years as the force law of attraction is another thing altogether. Familiarity breeds contempt in natural philosophy as in other contexts. There is always a danger in the unthinking repetition of the familiar.

In UFT 143 on www.aias.us the position tetrad was introduced and defined as:

$$R^a_\mu = R q^a_\mu \tag{13}$$

where R has the units of metres and where q^a_μ is the Cartan tetrad [1–10]. The linear velocity was worked out in condensed differential form notation as:

$$v^a = D \wedge R^a. \tag{14}$$

In tensor notation this becomes:

$$v^a_{\mu\nu} = c\left(\partial_\mu R^a_\nu - \partial_\nu R^a_\mu + \omega^a_{\mu b} R^b_\nu - \omega^a_{\nu b} R^b_\mu\right) \tag{15}$$

and in vector notation:

$$\mathbf{v}^a = \frac{\partial \mathbf{r}^a}{\partial t} + c\nabla R^a_0 + c\omega^a_{0b}\mathbf{R}^b - cR^b_0\boldsymbol{\omega}^a_b \tag{16}$$

where the omega symbol denotes the spin connection. The simplest type of antisymmetry law is now applied:

$$\frac{\partial \mathbf{R}^a}{\partial t} + c\omega^a_{0b}\mathbf{R}^b = -c\nabla R^a_0 + cR^b_0\boldsymbol{\omega}^a_b \tag{17}$$

choosing a diagonal spin connection:

$$a = b. \tag{18}$$

Then for each a:

$$\mathbf{v} = 2\left(\frac{\partial \mathbf{R}}{\partial t} + c\omega\mathbf{R}\right). \tag{19}$$

The factor 2 can be eliminated for ease of development and without loss of generality by defining:

$$\mathbf{r} = \frac{1}{2}\mathbf{R} \tag{20}$$

giving the simple expression:

$$\mathbf{v} = \left(\frac{\partial}{\partial t} + c\omega\right)\mathbf{r}. \tag{21}$$

The concept of proper time τ can be defined as:

$$\frac{\partial}{\partial \tau} = \frac{\partial}{\partial t} + c\omega. \tag{22}$$

The derivative appearing in Eq. (21) is the simplest type of covariant derivative in this theory.

There is nothing to guide this new theory in the mathematically incorrect Einstein general relativity, but the method of special relativity can still be used and developed. This was Einstein's original intention, which we seek to carry out in this paper.

The method used by Einstein to define the relativistic kinetic energy in special relativity [13] was based on classical dynamics, in which the work done is defined by:

$$W_{12} = T_2 - T_1 = \int_1^2 \mathbf{F} \cdot d\mathbf{r} \tag{23}$$

where T denotes the kinetic energy. Using the chain rule:

$$\mathbf{F} \cdot d\mathbf{r} = m\frac{d\mathbf{v}}{dt} \cdot \frac{d\mathbf{r}}{dt} dt = \frac{m}{2}\frac{d}{dt}(\mathbf{v} \cdot \mathbf{v})dt = d\left(\frac{1}{2}mv^2\right) \tag{24}$$

so

$$T_2 - T_1 = \int_1^2 d\left(\frac{1}{2}mv^2\right). \tag{25}$$

If the original kinetic energy is zero, the familiar expression for kinetic energy is recovered:

$$T = \frac{1}{2}mv^2. \tag{26}$$

In special relativity [13] the law of conservation of momentum demands that the momentum be defined as the relativistic momentum:

$$\mathbf{p} = \gamma m \mathbf{v} \tag{27}$$

where the Lorentz factor is the well known:

$$\gamma = \left(1 - \frac{v^2}{c^2}\right)^{-1/2} \tag{28}$$

in which $\mathbf{v}$ is the constant velocity of one frame with respect to the other and where c is the speed of light in the vacuum. Despite the fact that the momentum is worked out with the proper time, the force in Einstein's method [13] is worked out with the observer time t:

$$\mathbf{F} = d\mathbf{p} / dt \tag{29}$$

and the work integral is evaluated by integration by parts to give:

$$\begin{aligned} W = T &= \int \frac{d}{dt}(\gamma m \mathbf{v}) \cdot \mathbf{v} dt = m \int_0^v v d(\gamma v) \\ &= \gamma m v^2 - m \int_0^v \frac{v dv}{\left(1 - v^2 / c^2\right)^{1/2}} \\ &= \gamma m v^2 + mc^2 \left(1 - v^2 / c^2\right)^{1/2} - mc^2. \end{aligned} \tag{30}$$

The kinetic energy in special relativity is therefore:

$$T = (\gamma - 1) mc^2. \tag{31}$$

The total energy in special relativity is [13]:

$$E = \gamma mc^2 \tag{32}$$

and the rest energy is:

$$E_0 = mc^2. \tag{33}$$

So the relativistic kinetic energy is:

$$T = E - E_0. \tag{34}$$

These ideas are consequences of the Lorentz transformation, and what is really tested experimentally is the difference between proper time and the observer time.

As shown in earlier papers of this series, the ideas of the de Broglie Einstein theory of particle collisions have disintegrated completely, even within the context of special relativity. This may be due to this fundamental self inconsistency in Einstein's definition of force. These are papers published this year in Ref. [1].

The first step towards a new relativity is to define the kinetic energy from Eqs. (21), (23) and (29). In the first instance the method used by Einstein is followed for the sake of argument only. The fully consistent method would define the acceleration from the velocity using the spin connection. As in UFT 143 (www.aias.us) this method produces new types of acceleration and force in dynamics. Therefore, accepting Eq. (29) for the sake of argument, the kinetic energy in this new theory is:

$$\begin{aligned} T &= \frac{1}{2}mv^2 + mc\int \omega \mathbf{v}\cdot d\mathbf{r} \\ &= \frac{1}{2}mv^2 + mc\int \omega \frac{d\mathbf{v}}{dt}\cdot d\mathbf{r}\,dt \\ &= \frac{1}{2}mv^2 + mc\iint \omega \frac{d\mathbf{v}}{dt}\cdot\frac{d\mathbf{r}}{dt}\,dt\,dt \\ &= \frac{1}{2}mv^2 + c\iint \omega d\left(\frac{1}{2}mv^2\right)dt \\ &= \frac{1}{2}mv^2\left(1 + c\int \omega dt\right) \end{aligned} \tag{35}$$

Assume that the integral in Eq. (35) can be carried out from an initial time:

$$t = 0 \tag{36}$$

to a final time defined by t_f. The time t_f is a constant for a given cosmological system, and characteristic of that system. So the kinetic energy is:

$$T = \frac{1}{2}mv^2\left(1 + c\omega t_f\right). \tag{37}$$

The lagrangian of the system is:

$$\mathcal{L} = T - V = \frac{1}{2}mv^2\left(1 + \omega c t_f\right) - V \tag{38}$$

and its hamiltonian is:

$$H = T + V = \frac{1}{2}mv^2\left(1 + \omega c t_f\right) + V \tag{39}$$

where V is the potential energy of the system.

In cylindrical polar coordinates the relativistic lagrangian is:

$$\mathcal{L}=\frac{1}{2}m\left(\dot{r}^{2}+r^{2}\dot{\theta}^{2}\right)\left(1+\omega ct_{f}\right)-V\left(r\right) \tag{40}$$

The two Euler–Lagrange equations are [13]:

$$\frac{\partial \mathcal{L}}{\partial r}=\frac{d}{dt}\frac{\partial \mathcal{L}}{\partial \dot{r}}, \tag{41}$$

and

$$\frac{\partial \mathcal{L}}{\partial \theta}=\frac{d}{dt}\frac{\partial \mathcal{L}}{\partial \dot{\theta}}. \tag{42}$$

The radial tetrad (13) is expressed in the frame of the observer, and so is the velocity vector (21) and kinetic energy (35). The potential energy (V) is also expressed in the frame of the observer. Therefore, the Euler Lagrange equations (41) and (42) are also expressed in the observer frame. The spin connection term in the kinetic energy changes the theory into a relativistic theory on the basic assumption that physics is geometry, and that a spacetime with a connection is needed for relativity in general, i.e. general relativity with accelerations and forces.

Physics is currently entering an era of deep uncertainty after the complete collapse of Einstein general relativity, Einstein–deBroglie theory in special relativity, Higgs boson theory and string theory, so all assumptions must be questioned.

From Eq. (41), the force is:

$$F\left(r\right)=\left(m\ddot{r}-mr\dot{\theta}^{2}\right)\left(1+\omega ct_{f}\right)-\frac{1}{2}mct_{f}\frac{\partial\omega}{\partial r}\left(\dot{\theta}^{2}r^{2}+\dot{r}^{2}\right) \tag{43}$$

and as the spin connection vanishes it reduces to the Newtonian result [13]:

$$F\left(r\right)=m\ddot{r}-mr\dot{\theta}^{2}. \tag{44}$$

Eq. (43) is the required force law for any orbit. From Eq. (42) the constant total angular momentum is:

$$L=\frac{\partial \mathcal{L}}{\partial \dot{\theta}}=mr^{2}\left(1+\omega ct_{f}\right)\dot{\theta}. \tag{45}$$

The force law can be put into a more convenient format by defining:

$$u=1/r \tag{46}$$

so:

$$\frac{du}{d\theta} = -\frac{1}{r^2}\frac{dr}{d\theta} = -\frac{1}{r^2}\frac{dr}{dt}\frac{dt}{d\theta}. \tag{47}$$

From Eq. (45):

$$\frac{dt}{d\theta} = \frac{mr^2}{L}\left(1 + \omega ct_f\right) \tag{48}$$

so:

$$\frac{du}{d\theta} = -\frac{m}{L}\left(1 + \omega ct_f\right)\frac{dr}{dt} \tag{49}$$

and

$$\dot{r} = -\frac{L}{m\left(1 + \omega ct_f\right)}\frac{du}{d\theta}. \tag{50}$$

Next evaluate:

$$\frac{d^2u}{d\theta^2} = -\left(1 + \omega ct_f\right)\frac{m}{L}\frac{d}{d\theta}\left(\frac{dr}{dt}\right) \tag{51}$$

where:

$$\frac{d}{d\theta} = \frac{dt}{d\theta}\frac{d}{dt}. \tag{52}$$

So:

$$\frac{d^2u}{d\theta^2} = -\left(1 + \omega ct_f\right)\frac{m}{L}\frac{dt}{d\theta}\frac{d^2r}{dt^2}. \tag{53}$$

From Eq. (48):

$$\frac{d^2u}{d\theta^2} = -\left(1 + \omega ct_f\right)^2\left(\frac{m}{L}\right)^2 r^2\frac{d^2r}{dt^2} \tag{54}$$

so:

$$\ddot{r} = -\frac{L^2u^2}{m^2\left(1 + \omega ct_f\right)^2}\frac{d^2u}{d\theta^2}. \tag{55}$$

From Eq. (45):

$$r\dot{\theta}^2 = \frac{L^2 u^3}{m^2 \left(1+\omega c t_f\right)^2}. \tag{56}$$

Therefore the force law is:

$$\begin{aligned} -\frac{mF(u)}{L^2 u^2} &= \left(1+\omega c t_f\right)^{-1} \frac{d^2 u}{d\theta^2} + u \\ &+ \frac{1}{2}\frac{c t_f}{u^2}\frac{\partial \omega}{\partial r}\left(1+\omega c t_f\right)^{-2}\left(u^2 + \left(\frac{du}{d\theta}\right)^2\right) \end{aligned} \tag{57}$$

The force F may be found form a parameterization of any orbit:

$$u = \frac{1}{r} = f(\theta). \tag{58}$$

The orbit is therefore described using $\omega, \partial\omega / \partial r$, and the constant t_f In the Newtonian limit:

$$\omega \to 0 \tag{59}$$

and Eq. (57) reduces to

$$-\frac{mF(u)}{d^2\theta^2} = \frac{d^2 u}{d\theta^2} + u \tag{60}$$

This equation shows that for the ellipse:

$$r = \frac{\alpha}{1+\epsilon\cos\theta} \tag{61}$$

the force law is the familiar:

$$F(r) = -\frac{mMG}{r^2}. \tag{62}$$

It has been shown that it is possible to describe cosmology in terms of Cartan geometry and a unified field theory, ECE theory. With a given model for the spin connection, the characteristic time t_f can be found.

3. Classical limit and concept of force

Consider the elliptical orbit:

$$r = \frac{\alpha}{1+\epsilon\cos\theta} \tag{63}$$

and the unit vectors $\mathbf{e}_r$ and $\mathbf{e}_\theta$ of the cylindrical polar system. As in ref. [13]:

$$\dot{\mathbf{e}}_r = \dot{\theta}\mathbf{e}_\theta,\ \ \dot{\mathbf{e}}_\theta = -\dot{\theta}e_r. \tag{64}$$

The linear velocity is:

$$\mathbf{v} = \dot{\mathbf{r}} = \frac{dr}{dt} = \frac{d}{dt}(r\mathbf{e}_r) = \dot{r}\mathbf{e}_r + r\dot{\mathbf{e}}_r \tag{65}$$

and the linear acceleration is:

$$\mathbf{a} = \frac{d\mathbf{v}}{dt} = \left(\ddot{r} - r\dot{\theta}^2\right)\mathbf{e}_r + \left(r\ddot{\theta} + 2\dot{r}\dot{\theta}\right)\mathbf{e}_\theta \tag{66}$$

The calculation takes place in the frame of the observer so the time t is used. The elliptical orbit is observed in this frame. The force on an object of mass m is:

$$\mathbf{F} = m\mathbf{a}. \tag{67}$$

This object orbits an object of mass M, which in the solar system is the sun. The unit vectors of the cylindrical polar coordinates in the plane of the orbit are related to the Cartesian unit vectors by [14]:

$$\mathbf{e}_r = \mathbf{i}\cos\theta + \mathbf{j}\sin\theta \tag{68}$$

$$\mathbf{e}_\theta = -\mathbf{i}\sin\theta + \mathbf{j}\cos\theta. \tag{69}$$

Using the Lagrangian method [13] the total angular momentum of the system is:

$$L = mr^2\dot{\theta} = mr^2\frac{d\theta}{dt} \tag{70}$$

and is a constant of motion. From Eq. (63):

$$\frac{dr}{d\theta} = \frac{\epsilon}{\alpha}r^2\sin\theta \tag{71}$$

so:

$$\frac{dr}{dt} = \frac{dr}{d\theta}\frac{d\theta}{dt} = \left(\frac{L\epsilon}{m\alpha}\right)\sin\theta \tag{72}$$

i.e.:

$$\dot{r} = \left(\frac{L\epsilon}{m\alpha}\right)\sin\theta,\ \dot{\theta} = \frac{L}{mr^2}. \tag{73}$$

Therefore:

$$\ddot{r} = \left(\frac{L\epsilon}{m\alpha}\right)\frac{d}{dt}(\sin\theta),\ \ddot{\theta} = \frac{L}{m}\frac{d}{dt}\left(\frac{1}{r^2}\right). \tag{74}$$

Now use the chain rules:

$$\frac{df(r)}{d\theta} = \frac{df(r)}{dr}\frac{dr}{d\theta}, \tag{75}$$

$$\frac{df(\theta)}{dr} = \frac{df(\theta)}{d\theta}\frac{d\theta}{dr}, \tag{76}$$

to find:

$$\frac{d}{dt}\frac{1}{r^2} = -\frac{2}{r^3}\frac{dr}{dt},\ \frac{d}{dt}(\sin\theta) = \cos\theta\frac{d\theta}{dt}. \tag{77}$$

Therefore:

$$\ddot{r} = \left(\frac{\epsilon L^2}{m^2\alpha}\right)\frac{1}{r^2}\cos\theta,\ \ddot{\theta} = -\left(\frac{2L^2\epsilon}{m^2\alpha}\right)\frac{\sin\theta}{r^3}. \tag{78}$$

It follows that:

$$\ddot{r} - r\dot{\theta}^2 = \frac{\epsilon MG}{r^2}\cos\theta - \frac{L^2}{m^2r^3} \tag{79}$$

where G is Newton's constant. We have used [13]:

$$\alpha = \frac{L^2}{m^2MG}. \tag{80}$$

Similarly:

$$r\ddot{\theta}+2\dot{r}\dot{\theta}=0. \tag{81}$$

So the force is:

$$\mathbf{F}=m\left(\frac{\epsilon MG}{r^2}\cos\theta-\frac{L^2}{m^2r^3}\right)\mathbf{e}_r \tag{82}$$

and is radially directed.

Now use Eq. (63) again to find:

$$\epsilon\cos\theta=\frac{\alpha}{r}-1 \tag{83}$$

so the force is:

$$\mathbf{F}=\left(-\frac{mMG}{r^2}+\frac{L^2}{mr^3}-\frac{L^2}{mr^3}\right)\mathbf{e}_r \tag{84}$$

and is the familiar inverse square law of Robert Hooke [15] attributed to Isaac Newton by physicists. The latter define the traditionally named 'centrifugal force' as:

$$\mathbf{F}_c=\frac{L^2}{mr^3}\mathbf{e}_r \tag{85}$$

so the force in Eq. (84) is:

$$\mathbf{F}=\left(-\frac{mMG}{r^2}+F_c-F_c\right)\mathbf{e}_r \tag{86}$$

and is due to a precise cancellation of the traditionally named centrifugal force. The only thing that has been used in this calculation is the definition of the cylindrical polar coordinate system in a plane [13]. For a circular orbit:

$$\epsilon=0 \tag{87}$$

and

$$r=\alpha. \tag{88}$$

In this case, the force is an inverse cube in r:

$$\mathbf{F} = -\frac{L^2}{mr^3}\mathbf{e}_r \tag{89}$$

and agrees precisely with that obtained in UFT 193 (www.aias.us) by the lagrangian method. It is seen that the force for a circular orbit is exactly the negative of the traditionally named centrifugal force.

In an orbit the net force on m is zero in traditional understanding or received opinion. In the latter, the force is defined by the potential energy of the hamiltonian:

$$H = T + V \tag{90}$$

i.e.

$$F = -\frac{\partial V}{\partial r} \tag{91}$$

The potential energy U is defined as m multiplied by the gravitational potential:

$$V = m\Phi. \tag{92}$$

In order to obtain an inverse square law (84) the potential must be:

$$V = \frac{-mMG}{r}. \tag{93}$$

in the received opinion. However, from Eq. (84), the complete potential must be:

$$V = -\frac{mMG}{r} + \frac{L^2}{2mr^2} - \frac{L^2}{2mr^2}. \tag{94}$$

The received opinion gives an incomplete view of an orbit and force law.

What is observed experimentally in the solar system is the precessing ellipse:

$$r = \frac{\alpha}{1 + \epsilon\cos(x\theta)} \tag{95}$$

where x is the precession constant. In the received opinion of physicists the precession is due to the incorrect Einstein general relativity, but in Section 2 this view has been corrected and a new definition introduced of relativistic kinetic energy using the spin connection of Cartan. In the solar system x differs from unity only in the fifth or sixth decimal place, so the planetary precession is tiny, a few arc seconds per century. To an excellent approximation therefore the kinetic energy of Section 2 reduces to:

$$T = \frac{1}{2} m v^2. \tag{96}$$

The lagrangian is:

$$\mathcal{L} = T - V \tag{97}$$

and the total angular momentum is [13]:

$$L = \frac{\partial \mathcal{L}}{\partial \dot{\theta}} = m r^2 \dot{\theta}. \tag{98}$$

This angular momentum is a constant of motion and can also be obtained from:

$$\mathbf{L} = \mathbf{r} \times \mathbf{p}. \tag{99}$$

In this approximation:

$$\dot{r} = \left(\frac{x L \epsilon}{m \alpha} \right) \sin(x\theta), \ \dot{\theta} = \frac{L}{m r^2} \tag{100}$$

and it follows that:

$$\begin{aligned} \ddot{r} &= \frac{x L \epsilon}{m \alpha} \frac{d}{dt} \left(\sin(x\theta) \right) \\ &= \frac{x L \epsilon}{m \alpha} \frac{d}{d\theta} \left(\sin(x\theta) \right) \frac{d\theta}{dt} = \frac{x^2 L \epsilon}{m \alpha r^2} \cos(x\theta) \end{aligned} \tag{101}$$

and

$$\ddot{\theta} = -\frac{2 L^2 x \epsilon}{m^2 \alpha} \frac{\sin(x\theta)}{r^3}. \tag{102}$$

Therefore:

$$\ddot{r} - r\dot{\theta}^2 = \frac{\epsilon x^2 M G}{r^2} \cos(x\theta) - \frac{L^2}{m^2 r^3}, \tag{103}$$

$$r\ddot{\theta} + 2\dot{r}\dot{\theta} = 0. \tag{104}$$

Using Eq. (66) for linear acceleration and Eq. (67) for the force:

$$\mathbf{F} = m\ddot{\mathbf{a}} = m\left(\frac{x^2\epsilon MG}{r^2}\cos\left(x\theta\right)-\frac{L^2}{m^2r^3}\right)\mathbf{e}_r. \tag{105}$$

Finally, use Eq. (95):

$$\epsilon\cos\left(x\theta\right)=\frac{\alpha}{r}-1 \tag{106}$$

and Eq. (80), which is true in the solar system to an excellent approximation, to find that:

$$\mathbf{F}=\left(-\frac{x^2mMG}{r^2}+\frac{L^2}{mr^3}\left(x^2-1\right)\right)\mathbf{e}_r \tag{107}$$

which is exactly the same as the result obtained in Eq. (9) of UFT 193 (www.aias.us).

So for tiny precessions of the perihelion, the force law is a combination of an inverse square and inverse cube in r. The Einstein theory is incorrect, because using the same lagrangian method it gives a sum of inverse square and inverse fourth terms. By direct checking with computer algebra in UFT 193, the sum used by Einstein general relativity does not give a precessing ellipse. The Einstein theory is now known clearly to scientists to be incorrect for many other reasons. The fully relativistic theory of the precessing elliptical orbit is given in Section 2, using the Cartan spin connection.

The calculations in this Section 3 also have implications in the traditional classical dynamics known to physicists as Newtonian dynamics. They start with the analytical function of the ellipse in cylindrical polar coordinates and give the force (86) using the definition of derivatives in the cylindrical polar coordinates, and nothing else. No other concept is used. The force law (86) contains an exact cancellation of the term known traditionally to physicists as 'the centrifugal force of repulsion'. This idea has always given a lot of trouble, and by inspection of Eq. (84) it becomes clear that it appears from the orbital equation but cancels out to give the force of attraction known as the inverse square law of attraction. The use of the cylindrical polar coordinate system gives a sum of the mathematical object known as the centrifugal force. One term of the sum is positive, the other term is negative. For a static ellipse (Eq. (63)) the two terms cancel exactly, but for a precessing ellipse (Eq. (95)) they do not, and give the inverse cube term. The origin of the centrifugal term is the movement in the cylindrical polar system of the unit vectors, which are not necessarily constant in time. This fundamentally important point is discussed in chapter one of ref. [13] and is true independently of the Newton laws of motion as they are known traditionally in the dogma of physicists. In historical fact Newton inferred only the third law. In Newtonian dynamics the precise origin of the 'centrifugal force' is the angular part of the kinetic energy in cylindrical polar coordinates:

$$T = \frac{1}{2} m r^2 \dot{\theta}^2 = \frac{L^2}{2mr^2}. \tag{108}$$

The total energy is written as [13]:

$$E = \frac{1}{2} m \dot{r}^2 + \frac{L^2}{2mr^2} + V(r) \tag{109}$$

and the received opinion asserts that the effective potential energy is [13]:

$$u(r) = V(r) + \frac{L^2}{2mr^2} \tag{110}$$

giving rise to the received opinion of an orbit as the balance of an attractive, negative valued part of the 'effective' potential (110) and a positive valued repulsive part, the centrifugal potential energy. This is obviously incorrect dogma, because the object known as a centrifugal potential of repulsion is in fact part of the kinetic energy. There is no potential of repulsion in Newtonian dynamics, which does not therefore describe orbits at all. It merely describes the radially directed force of attraction between two objects. It is concluded that the object known to physicists as 'force' in Eq. (86) is a re-formulation of the analytical equation of an ellipse, and no more than that. It is not a 'universal' force of gravitation as claimed in the dogma. It varies from orbit to orbit.

The fully correct description of orbits requires the spin connection in ECE theory, in which the centripetal force is part of the spacetime torsion in papers such as UFT 55 (www.aias.us), dealing with the non-inertial Coriolis dynamics.

References

[1] M .W. Evans, Ed., Journal of Foundations of Physics and Chemistry, Cambridge International Science Publishing (CISP), (www.cisp-publishing.com).

[2] M. W. Evans, S. Crothers, H. Eckardt and K. Pendergast, "Criticisms of the Einstein Field Equation" (CISP, 2011).

[3] K. Pendergast, "The Life of Myron Evans" (CISP, 2011).

[4] M .W. Evans, H. Eckardt and D. Lindstrom, "Generally Covariant Unified Field Theory" (Abramis Academic, 2005 to 2011), in seven volumes.

[5] L. Felker, "The Evans Equations of Unified Field Theory" (Abramis 2007).

[6] The ECE websites: www.aias.us, www.webarchive.org.uk, www.atomicprecision.com, www.et3m.net, www.upitec.org.

[7] M. W. Evans, H. Eckardt and D. W. Lindstrom, Papers and plenary published by the Serbian Academy of Sciences, 2010 and 1011.

[8] M .W. Evans and S. Kielich (eds.), "Modern Nonlinear Optics" (Wiley, 1992, 1993, 1997 and 2001), first and second editions in six volumes.

[9] M .W. Evans and L. B. Crowell, "Classical and Quantum Electrodynamics and the B(3) Field" (World Scientific, 2001).

[10] M. W. Evans and J.-P. Vigier, "The Enigmatic Photon" (Kluwer 1994 to 2002), in five volumes.

[11] A. A. Vankov, www.wbabin.net/eeuro/vankov.pdf

[12] S. P. Carroll, "Spacetime and Geometry: an Introduction to General Relativity" (Addison Wesley, New York, 2004).

[13] J. B. Marion and S. T. Thornton, "Classical Dynamics of Particles and Systems, Third Edition" (HBJ Publishing, 1988).

[14] E. G. Milewski (Chief Ed.), "Vector Analysis Problem Solver" (Research and Education Association, New York, Revised Printing, 1987).

[15] John Aubrey, "Brief Lives" (from a seventeenth century collection in the Bodleian Library Oxford, literary classic).

A simple explanation of the whirlpool galaxy from the new ECE relativity

M.W. Evans[1*] and H. Eckardt[2]**

**Alpha Institute for Advanced Studies (AIAS) (www.aias.us); **Unified Physics Institute of Technology (UPITEC) (www.upitec.org)*

The velocity curve of a whirlpool galaxy is described by the new ECE relativity developed in recent papers of this series, papers which have refuted the Einsteinian general relativity. The orbital linear velocity is defined by the cylindrical polar coordinates in a plane, and the constant total angular momentum of the system defined in terms of the spin connection and a characteristic time. It is shown that all orbits can be described by the relevant spin connection magnitude, so this is a new theory of general relativity that can be applied to astronomical observations without the use of dark matter and without the Einstein theory. Therefore this is a new relativistic explanation without the use of dark matter.

Keywords: New ECE relativity, velocity curve of a whirlpool galaxy, rotational Hooke law.

1 Introduction

In recent papers of this series of one hundred and ninety seven papers to date [1-10] Einsteinian general relativity has been thoroughly refuted in several ways. This means that a century of standard model relativity is meaningless, both theoretical and experimental, and is an example of Langmuir's pathological science, or carelessly repeated dogma coming from an unwillingness or inability to test the original concepts with sufficient mathematical rigour. Astonishingly, the first refutation of Einstein's general relativity occurred barely a month after he had published a paper on the perihelion advance of Mercury. The paper was published in November 1915 in the Proceedings of the Royal Prussian Academy, and has been translated recently and heavily criticised on the web for basic mathematical errors that nullify its physics [11]. Reference (11) translates a letter sent to Einstein by Karl Schwarzschild in December 1915. The letter refutes Einstein's paper comprehensively, and introduces the genuine, or real and original Schwarzschild metric. Einstein published only one minor note on this subject after

[1]e-mail: EMyrone@aol.com
[2]e-mail: horsteck@aol.com

that. In a later fabrication a different metric was attributed to Schwarzschild after his death in 1916 of disease in the trenches. This fabrication became known as "the Schwarzschild metric". In recent papers UFT190 ff. of this series (www.aias.us) this fabrication has been tested directly using computer algebra and has been shown to be total nonsense. For example it does not produce a precessing elliptical orbit as claimed so often by the dogmatists. In effect, this is exactly what was pointed out by Schwarzschild to Einstein by letter in December 1915, nearly a hundred years ago.

So Einstein's general relativity should have been abandoned in December 1915. If Schwarzschild had lived he would probably have out argued Einstein before the latter's dogma started to balloon out of control. Schroedinger also started to criticize Einstein in 1918, along with Bauer. Unfortunately Eddington started to claim that his experiments supported the light bending theory of Einstein, and the latter's dogma was blown out of all proportions by a media that did not understand it, the twentieth century disease. In UFT 150B and 155 of this series (www.aias.us) and in a series of essays, the general public has been reeducated and now know that the light bending theory is also riddled with errors which nullify its physics. In one of the supreme ironies of physics, the magnitude of the light bending is known with precision at tremendous cost, and attributed to complete mathematical nonsense. Much worse even than this for standard physics is the incorrectness of the Einstein field equation due to neglect of torsion, and again this nullifies all the physics of the twentieth century based on that field equation, fantasises such as big bang, black holes, dark matter and all that. Down the years others have suspected Einstein of promoting errors, or keeping quiet about them, and included Dirac, Eddington (despite his initial enthusiasm), the mathematician Levi-Civita who inferred curvature, Vigier (who came to reject general relativity in his last papers), and many others. Hoyle dismissed big bang out of hand and left Cambridge.

In the aftermath of this fiasco UFT 196 began the search for a new ECE relativity based on simplicity and Ockham's Razor, rejecting the meaningless and mathematically incorrect complexity of twentieth century relativity. This search was initiated using the concept of the velocity tetrad [1 - 10]. As a first step into unknown territory, the methods of special relativity were adopted to calculate the relativistic kinetic energy and the correctly relativistic total angular momentum. In future work these initial steps ought to be looked at critically and developed but we intend to try to keep it simple - Ockham's Razor and as in this paper try to test it simply against experimental data - Baconian principles of science.

In Section 2 the orbital linear velocity is calculated in cylindrical polar coordinates in a plane and expressed in terms of the angular velocity. The latter is developed in terms of the constant total angular momentum, which is expressed in the simplest possible way in terms of the Cartan spin connection and a characteristic time of the system. This time interval is a constant akin to the idea of relaxation time and originates in the need for a fully relativistic kinetic energy. Line element general relativity has been refuted trivially for all spherical spacetimes in UFT 194, and can no longer be used, so a new method of calculating relativistic quantities has had to be introduced. Using this definition of the constant total angular momentum, a constant of motion, the orbital linear velocity can be expressed in terms of the orbital function, found by observation in astronomy. It is shown in Section 2 that the observed velocity curve of a whirlpool galaxy can be explained straightforwardly in terms of the spin connection magnitude for any type of spiral arrangement of stars. The orbital linear velocity reaches a plateau with increasing r, the distance between the star and the centre of the galaxy. Finally the fundamental relativistic explanation of the whirlpool galaxy is expressed as a driving torque. The torque originates in spacetime torsion, which is governed by Cartan's geometry and which gives rise to the ECE field equations of gravitation and electrodynamics in a generally covariant unified field theory [1 - 10] - ECE theory. For a hyperbolic spiral for example the torque is proportional the square of angular displacement.

In Section 3 the various spiral orbits considered in this paper are illustrated by computer together with the spin connections for each type of orbit. This method can be extended to any type of orbit observed in astronomy and gives a self consistent cosmology.

2 Explanation of the velocity curve

In cylindrical polar coordinates the orbital linear velocity of a star in a whirlpool galaxy is defined as [12]:

$$\mathbf{v} = \frac{dr}{dt}\mathbf{e}_r + \frac{r\,d\theta}{dt}\mathbf{e}_\theta = \frac{d\theta}{dt}\left(\frac{dr}{d\theta}\mathbf{e}_r + r\mathbf{e}_\theta\right). \tag{1}$$

In UFT 196 (www.aias.us) the constant total angular momentum of the system was shown to be:

$$L = mr^2\left(1+\omega ct_f\right)\frac{d\theta}{dt} \tag{2}$$

where m is the mass of the star, r the distance between the star and the centre of the galaxy, the magnitude of the spin connection and t_f a characteristic and constant interval of time of the system. This is a fully relativistic description based on Cartan's geometry. Therefore the angular velocity of the star is:

$$\frac{d\theta}{dt} = \frac{L}{mr^2(1+\omega ct_f)} \tag{3}$$

about the centre of the galaxy. The velocity can therefore be expressed as:

$$\mathbf{v} = \frac{L}{mr^2(1+\omega ct_f)}\left(\frac{dr}{d\theta}\mathbf{e}_r + r\mathbf{e}_\theta\right) \tag{4}$$

and the square of the velocity is:

$$\mathbf{v}^2 = \left(\frac{L}{mr^2(1+\omega ct_f)}\right)^2\left(\left(\frac{dr}{d\theta}\right)^2 + r^2\right). \tag{5}$$

In a whirlpool galaxy it is observed that v becomes constant with increasing r, and this velocity curve cannot be described by Einsteinian general relativity. The velocity curve is therefore described by:

$$v \underset{r\to\infty}{\longrightarrow} \text{constant} \tag{6}$$

From Eq. (5)

$$\left(\frac{dr}{d\theta}\right)^2 + r^2 = Ar^4\left(1+\omega ct_f\right)^2 \tag{7}$$

where:

$$A = \left(\frac{vm}{L}\right)^2. \tag{8}$$

So the magnitude of the spin connection is defined by:

$$\omega = \frac{1}{ct_f A^{1/2} r^2}\left(\left(\frac{dr}{d\theta}\right)^2 + r^2\right)^{1/2} - \frac{1}{ct_f} \tag{9}$$

for any orbital function $dr/d\theta$. Here t_f is a characteristic and constant time interval. The arrangement of the stars in a whirlpool galaxy is observed to be a spiral. It is shown as follows that the velocity curve can be explained in the new relativity irrespective of the type of spiral observed astronomically.

The hyperbolic spiral is given by:

$$\left(\frac{dr}{d\theta}\right)^2 = \frac{r^4}{r_0^2} \tag{10}$$

so the magnitude of the spin connection is:

$$\omega = \frac{1}{ct_f A^{1/2} r}\left(1+\left(\frac{r}{r_0}\right)^2\right)^{1/2} = \frac{1}{ct_f} \tag{11}$$

with the limiting property:

$$\omega \underset{r\to\infty}{\rightarrow} \frac{1}{ct_f A^{1/2} r_0} - \frac{1}{ct_f}. \tag{12}$$

So as r becomes infinite the spin connection for the hyperbolic spiral becomes constant because:

$$A \underset{r\to\infty}{\rightarrow} \text{constant} \tag{13}$$

i.e. v becomes constant as r approaches infinity.

The logarithmic spiral is defined by:

$$r = r_0 e^{\beta\theta} \tag{14}$$

where β is the pitch. So:

$$\frac{dr}{d\theta} = \beta r \tag{15}$$

and the magnitude of the spin connection is:

$$\omega = \frac{L}{ct_f vmr}\left(1+\beta^2\right)^{1/2} - \frac{1}{ct_f}. \tag{16}$$

Its limiting behaviour is:

$$\omega \underset{r\to\infty}{\rightarrow} \text{constant} \tag{17}$$

and as r becomes infinite both v and ω become constant.

The Archimedes spiral is given by:

$$r = a + b\theta \tag{18}$$

so its spin connection is:

$$\omega = \frac{1}{ct_f A^{1/2} r^2}\left(b^2 + r^2\right)^{1/2} - \frac{1}{ct_f} \tag{19}$$

with limiting behaviour:

$$\omega \underset{\to\infty}{\to} -\frac{1}{ct_f}. \tag{20}$$

So for the Archimedes spiral the spin connection becomes constant as r becomes infinite.

Fermat's spiral is given by:

$$r = r_0 \theta^{1/2} \tag{21}$$

so:

$$\theta = \left(\frac{r}{r_0}\right)^2 \tag{22}$$

and

$$\left(\frac{dr}{d\theta}\right)^2 = \frac{r_0^4}{4r^2}. \tag{23}$$

Its spin connection is therefore:

$$\omega = \frac{1}{ct_f A^{1/2} r^2}\left(\frac{r_0^4}{4r^2} + r^2\right)^{1/2} - \frac{1}{ct_f} \tag{24}$$

with the limiting behaviour

$$\omega \underset{r\to\infty}{\to} \frac{-1}{ct_f} \tag{25}$$

the same result as for the Archimedes spiral.

The lituus is defined by:

$$\theta = \left(\frac{r_0}{r}\right)^2 \tag{26}$$

so

$$\left(\frac{dr}{d\theta}\right)^2 = \frac{r^6}{4r_0^4} \tag{27}$$

and its spin connection is:

$$\omega = \frac{1}{ct_f A^{1/2} r}\left(1+\left(\frac{r}{r_0}\right)^4\right)^{1/2} - \frac{1}{ct_f}. \tag{28}$$

Its spin connection becomes infinite with infinite r.

There are many other types of spiral, usually the stars are fitted experimentally with a hyperbolic spiral. There is a spin connection for any type of spiral, or indeed any orbit.

In the solar system the astronomically observed orbital function is the precessing ellipse:

$$r = \frac{\alpha}{1+\epsilon\cos(x\theta)} \tag{29}$$

where 2α is the right latitude, ϵ the eccentricity, and x the precession constant. Its spin connection is:

$$\omega = \frac{L}{ct_f vmr}\left[1+\left(\frac{\epsilon xr}{\alpha}\right)^2\left(1-\frac{1}{\epsilon^2}\left(\frac{\alpha}{r}-1\right)^2\right)\right]^{1/2} = \frac{1}{ct_f} \tag{30}$$

The precessing ellipse reduces to a circle if

$$x = 1,\ \epsilon = 0,\ \alpha = r \tag{31}$$

in which case the spin connection becomes:

$$\omega = \frac{L}{ct_f vmr} - \frac{1}{ct_f} \tag{32}$$

and is a constant because for a circle:

$$L = m\ v\ r. \tag{33}$$

The earth's orbit is very nearly a circle, so its spin connection magnitude (units of inverse metres) is determined by the interval of time t_f. In the new general relativity the earth's orbit is due to the underlying spacetime torsion.

The torsion manifests itself as a torque magnitude:

$$Tq = L\frac{d\theta}{dt} = mr^2\left(1+\omega ct_f\right)\left(\frac{d\theta}{dt}\right)^2. \tag{34}$$

For each type of orbit there is a characteristic Tq. The hyperbolic spiral for example is given by:

$$Tq = \frac{L}{mr^2(1+\omega ct_f)} = k\theta^2 \tag{35}$$

where k is a constant. This is a driving torque, not a restoring torque, so the sign of k is positive. The torque is generated by the underlying spacetime torsion. The equation of the hyperbolic spiral is given from Eq. (35) as:

$$\theta = \left(\frac{L^2}{mk\left(1+\omega t_f\right)}\right)^{1/2}\frac{1}{r} \tag{36}$$

each type of orbit is driven by a torque generated by spacetime torsion. The latter is governed by Cartan geometry and the ECE field equations of dynamics and electrodynamics. One of the equations for dynamics reduces to the inverse square force law of attraction as the spin connection becomes zero. Usually this is attributed to Newton, but it was inferred originally by Hooke. In the received opinion the inverse square force law of attraction is negative valued and results in a static elliptical orbit. The received opinion takes the rotational kinetic energy energy and redefines it incorrectly as a centrifugal potential energy. A balance of the centrifugal force and attractive force is asserted to keep an object of mass m in an elliptical orbit.

In the new general relativity developed in this paper the orbit is the result of a fundamental spacetime geometry. For a static ellipse:

$$x = 1 \tag{37}$$

so the spin connection for a static elliptical orbit is defined by Eqs. (30) and (37). When there is no angular momentum:

$$L = 0 \tag{38}$$

there is no longer any rotational motion, and the spin connection vanishes. In this case the inverse square law of attraction is recovered through one of the ECE field equations. The latter are given by Cartan geometry. When there is no angular momentum there is a central force law of attraction between an object of mass m and one of mass M.

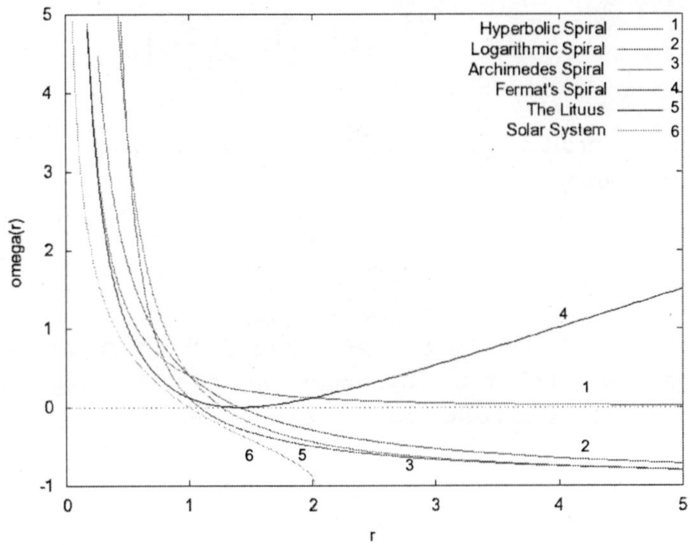

Fig. 1. Spin connection for various types of spiral orbits and the solar system.

3 Graphics

3.1 Spin connections

The radial development of the spin connections for the spiral orbits described in the preceding chapter has been plotted in Fig. 1. As can be seen, all spin connections converge to a constant value with exception of the Lituus which diverges for $r \to \infty$. All parameters in the spin connections have been set to unity with exception of $\epsilon = 0.1$ which occurs in the orbit of the solar system. The curve for the solar system ends at $r = 2$ where ω becomes complex valued. This is not astonishing since the limit $v =$ const for which the spin connections were derived is not true in the solar system. Furthermore, curve parts for small radius values are not reliable since there is a r dependence of the velocity curves in this regime.

3.2 Fine art graphics

Only so much can be deduced from a still, galactic photograph. That is, rotating the photograph does not simulate a rotating galaxy. Rather, the photograph is a snapshot of objects in spacetime that are frozen in their development. The snapshot is a replica of how the galaxy looks in some mid-development. So the image we have may or may not be a repeating cycle of symmetry as would appear in M51, the Whirlpool Galaxy with its symmetrical, spiral evolutions for example.

Current observations indicate that the spiral arms are sectioned by varying

Fig. 2. M51, the whirlpool galaxy - clockwise rotation.

Fig. 3. M74 – counter clockwire rotation

degrees of arc and have a generally short length to their "plume". Once joined in sequence and torsion however, they form the classic spiral development, appearing to be one long, spiralling body.

Nevertheless, observations also show a tendency in all spiral arms towards an overall, polygonal symmetry that manifests as skewed, semi hexagonal or semi-octagonal in presentation.

Closer scrutiny has also revealed a profusion of straight-line, trajectory traces that "cut" through the spiral developments and appear to have created interaction or collision with the more sedately orbiting and spiralling masses.

Further scrutiny implies that all matter begins orbit by degrees of vigorous radial trajectory and that there is visible evidence of further vigorous interaction, collision or inherent mass "explosion" traces peppering particularly, the whole of M51's galactic plane. In higher velocity, radial trajectory, it is inconceivable that masses or matter would not collide on a regular basis.

Energetic or explosive, bright white and red "bubbles" of activity, the bluish white type being reminiscent of a still-shot of a divers air bubble as it is caught shimmying to the surface, will also appear to be the instigators of more traces and tracks of ejected nebulae. It is worthy of note that in almost all cases, resultant nebulae and "blast debris" and its traces will invariably be on the outside only of the "bubbles" and outward bound, rarely if ever, on the core side or in that direction.

The apparent vigour of the outward blast should, in an inherent explosion, show 360 degrees of blast arc that may distort but would encroach inward at

Fig. 4. NGC 5584 – clockwise direction.

least to some degree. No such debris has been observed (Figs. 2–7).

The sum of observations thus far, imply vigorous ejection at the core of a radially inclined trajectory. The fastest of this matter achieves a straight-line trace after slight curvature near the core area and briefly on emanation only. The slowest of such matter will be in spiral trace orbit and develops a positive angular momentum of a minimal value which is proportional to the overall rotation speed.

Energetic activity or explosive collision in the middle reaches of the plane may form new plumes that mimic neighbouring, fully developed spiral arm orbits. This gives a false impression of more core generated arms than there actually are Figs. 5 & 7)

The true paths of these masses and matter are not accurately reflected in the photographic image. The Spiral Arm development is real and obvious enough. The radial traces, the "Cheshire Lines" are real enough if not so obvious. The missing, or more accurately, invisible half of the evolutionary picture is Torsion dominated. Doctors Evans and Eckardt have voiced a number of analogies that graphically describe the effects of torsion. Here I will refer to one of Dr. Evans' as it best describes the effects relevant to my present visual analysis:

A cylindrical glass rod is caused to rotate around its longer axis in a beaker of water (in practical terms and for ease of visualisation, by a pillar drill say, or similar) and without direct contact with the inside of the containing beaker.

Firstly, the water would begin to rotate en masse with and around the rod like a whirlpool. But for the friction between the base of the beaker and the table it sits on, the beaker would begin to spin in union. The same applies to the table

Fig. 5. M101 – counter clockwise direction.

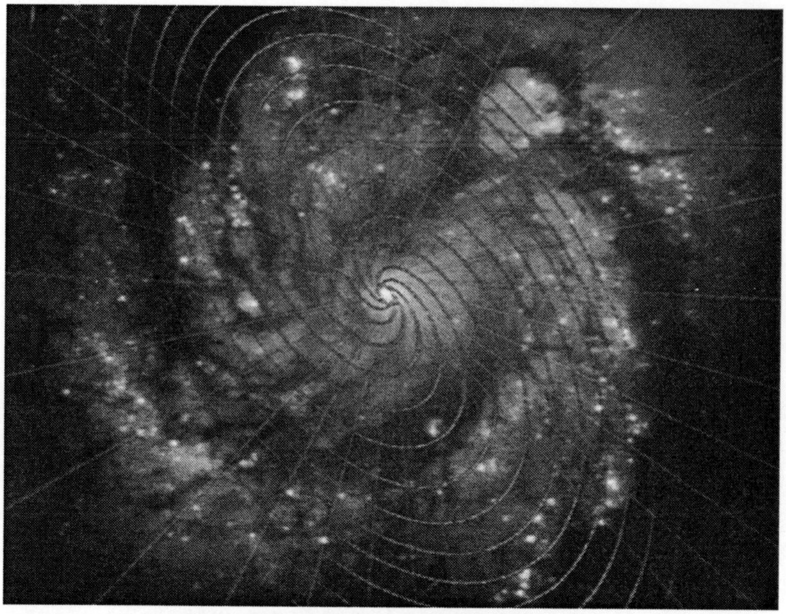

Fig. 6. M100 – clockwise direction.

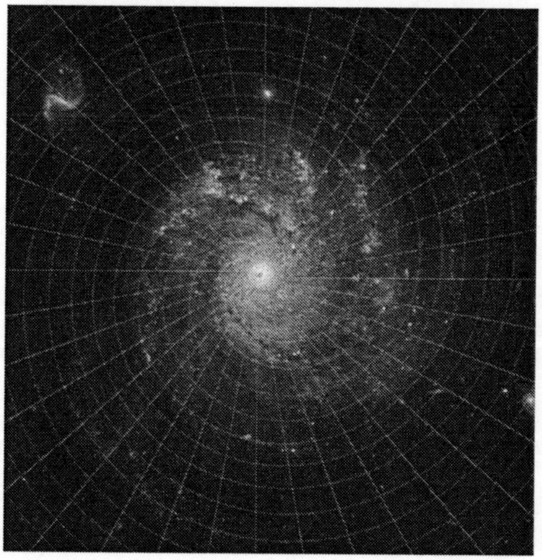

Fig. 7. NGC 1309 – counter clockwise direction.

and the floor, the floor and the building, the building and the ground it sits on and so on and so on. The "frictional drag" between the rod and the water, and that which draws in all things into the spin is caused by the Torsion in spacetime.

So a small, spinning mass will cause far reaching rotational effects in a much larger spacetime that surrounds it.

So it is in galaxies, moons, planets, suns , atoms, electrons and all spinning things, which is, to all intents and purposes, everything. Everything is spinning in spacetime.

With that picture firmly in mind, it can now be understood more easily that the galactic photograph is showing us relative motions only. The radial traces are very straight in the picture but the true orbital track of matter on this course in absolute space, will have a significant and more obvious curvature. I have deduced through animated simulation that in relation to one galactic revolution, the spiral arms will have a true, absolute spiral orbit path of twice the pitch of the visible arms.

For each revolution denoted by the relative spiral orbits in the galactic photograph, there has been an "invisible" orbit visually obscured by the revolution of the spacetime around the whole galaxy. Therefore, what appears in the photograph to be a spiral of specific angle is, without trickery and in reality a spiral of twice the angle which is identical to its visual product by scale of 100% . Only a factorial angle can do this. variation of angle here will change the visible spiral shape to angles which would be factors of that variation when related to one revolution.

References

[1] M. W. Evans, Ed., J. Foundations of Physics and Chemistry (Cambridge International Science Publishing, CISP, 2011 onwards, six issues a year, www.cisp-publishing.com).

[2] M .W. Evans, S. Crothers, H. Eckardt and K. Pendergast, "Criticisms of the Einstein Field Equation" (CISP, 2011).

[3] K. Pendergast, "The Life of Myron Evans" (CISP, 2011).

[4] M .W. Evans, H. Eckardt and D. W. Lindstrom, "Generally Covariant Unified Field Theory" (Abramis Academic, 2005 - 2011), in seven volumes.

[5] The ECE websites: www.aias.us (archived on www.webarchive.org.uk). www.atomicprecision.com, www.upitec.org, www.et3m.net.

[6] L. Felker, "The Evans Equations of Unified Field Theory" (Abramis 2007).

[7] M. W. Evans, H. Eckardt and D. W. Lindstrom, plenary and papers 2010 - 2011 of the Serbian Academy of Sciences.

[8] M. W. Evans and S. Kielich, "Modern Nonlinear Optics" (Wiley 1992, 1993, 1997,2001, in six volumes and two editions).

[9] M .W. Evans and L. B. Crowell, "Classical and Quantum Electrodynamics and the B(3) Field" (World Scientific 2001); M. W. Evans and A. A. Hasanein, "The Photomagneton in Quantum Field Theory" (World Scientific 1994).

[10] M. W. Evans and J.-P. Vigier, "The Enigmatic Photon" (Kluwer, 1994 to 2002), in ten volumes softback and hardback.

[11] A. A. Vankov, www.wbabin.net/eeuro/vankov.pdf.

[12] J. B. Marion and S. T. Thornton, "Classical Dynamics" (HBC, New York, 1988, $3^{rd.ed}$.).

ECE Theory of unified physics and refutations of the standard model

M.W. Evans[1*], H. Eckardt[2] and D.W. Lindstrom[3*]**

**Alpha Institute for Advanced Studies (AIAS) (www.aias.us); **Unified Physics Institute of Technology (UPITEC) (www.upitec.org)*

ABSTRACT: The progress is reviewed of papers 101 to 199 of the ECE theory of unified physics. The papers are classified as follows: 1) spin connection resonance (SCR) and computer simulation of spacetime devices; 2) geometry; 3) astronomy and cosmology; 4) precession phenomena; 5) conservation, continuity and Poynting theorems; 6) dynamics and fluid dynamics; 7) equivalence principles; 8) antisymmetry laws, refutations of classical electrodynamics; 9) fermion equation, quantum Hamilton, quantum force equations and quantum field theory, refutation of the Dirac equation; 10) photon mass and particle scattering theory, refutation of standard particle theory; 11) refutations of Einsteinian general relativity; 12) towards a new general relativity; 13) interaction of electromagnetism and gravitation and electrodynamics; 14) B(3) field theory; 15) geometrical refutations of the Einstein field equation and general relativity.

Keywords: ECE theory of unified physics, refutations of the standard model, review of papers 101 to 199 of ECE theory.

1. Introduction

There are one hundred and ninety nine papers available of the ECE theory of unified physics [1–10], and several other papers and volumes (www.aias.us). This article is a concise review of the major progress made in papers 101 to 199 of the series, the papers produced since review paper 100. The ECE theory is the new unified field theory of physics, and has replaced the obsolete standard model in several ways. The refutations of the standard model have been made side by side with the development of ECE theory, the well known acronym for Einstein Cartan Evans field theory. The central feature of ECE theory is that it unifies

[1]e-mail: EMyrone@aol.com
[2]e-mail: horsteck@aol.com
[3]e-mail: dwlindstrom@gmail.com

physics in terms of Cartan's well known geometry [11], the first geometry to emerge (in the nineteen twenties) that can be described as rigorously complete and self consistent in all mathematical spaces. The basic philosophy of ECE theory is that of Kepler and his renaissance contemporaries: ubi materia ibi geometria –all nature is geometry. This is also the basic principle of general relativity. One of the most important advances made by ECE theory in UFT 101 to UFT 199 on www.aias.us is the simple and conclusive refutation of Einstein's ideas about general relativity, and their replacement by a new relativity. Einstein's work is corrected and strengthened, his contributions remain a pinnacle of human thought: each generation stands on the shoulders of giants.

Similarly Dirac's equation has been refuted and replaced by the fermion equation of ECE theory, the Copenhagen interpretation of physics has been refuted and replaced by a Baconian quantum mechanics unified with general relativity. The great advances made by Dirac and others also remain a pinnacle of thought, and have been corrected and strengthened. Other major advances in ECE quantum mechanics include the inference and development of the quantum Hamilton and quantum force equations. The Maxwell–Heaviside (MH) equations of electrodynamics have been refuted by antisymmetry arguments of ECE and developed into general relativity as part of a unified field theory based on Cartan geometry. The conditions under which the well tested mathematical structure of the MH equations are recovered from the ECE theory have been defined with precision by Eckardt and Lindstrom [12–14]. The Einstein field equation has been refuted conclusively and thoroughly by antisymmetry arguments based on the ineluctable presence of torsion in a self consistent and complete geometry of the general mathematical space [1–10]. This means that the theory of big bang and of black holes has also been refuted, since these theories depend on an irretrievably incorrect Einstein field equation. The development and simplification of the geometry of ECE theory–Cartan geometry, has led to the engineering model of the theory, in which the equations of electrodynamics and dynamics have the same structure in the same, general, mathematical space. The dynamical equations of ECE include equations of the gravitomagnetic field, the equivalent of magnetic flux density in electrodynamics. It is inferred in ECE theory that each equation of electrodynamics has its dynamical counterpart. The ECE dynamical equations have been utilized to explain precessional phenomena in astronomy.

The basic features have been refuted of particle scattering theory in the obsolete standard physics. The Einstein de Broglie theory has been shown to be deeply self inconsistent. Standard Compton scattering theory has been shown to be applicable only when the photon mass is identically zero, a contradiction in thought that makes the standard model untenable at its most fundamental level. If the photon mass is finite, the standard theory of Compton scattering collapses completely, and it has been shown that general Compton scattering theory of two particles of mass also collapses into nonsense. Basic absorption theory and Raman scattering theory has been shown to be deeply self inconsistent. Astonishingly, neither considers conservation of momentum. When the latter and basic concept is introduced the standard theory becomes wildly untenable. This year, CERN has finally concluded that the Higgs boson does not exist, so another basic idea of the obsolete standard model has evaporated. During the evolution of ECE theory seen in UFT101 to UFT199 a new approach to scattering theory has been forged, known as R theory, and based on the ECE wave equation.

Metric based general relativity has been thoroughly refuted in a number of complementary ways, culminating in the discovery that there is no self consistent, metric based, general relativity in a spherical spacetime. It has been found that the only possible metric based general relativity must be based on the very general metric developed by Crothers [1–10]. In the series UFT 190 ff the metric based general relativity was refuted in several very simple ways, so there is no rational way of saving the metric based general relativity, the theory that is claimed incorrectly to account for light bending and time delay in the solar system. It has been shown that the theory of these phenomena due to Einstein contains several basic errors (UFT 150 and 155). There are many major consequences of these simple refutations, for example the "precision" tests of Einsteinian relativity in the solar system become obviously meaningless. Accurate historical scholarship during the course of development of ECE theory has shown that the Einstein theory of perihelion advance was dismissed as incorrect by Schwarzschild as early as December 1915, only a month after the theory was first published. In UFT 190 ff (www.aias.us) it has been shown that the metric falsely but commonly attributed to Schwarzschild is completely meaningless. This fact becomes crystal clear in at least two ways: firstly the false Schwarzschild metric does not produce a precessing elliptical orbit; secondly UFT 194 shows in a perfectly simple and general manner that the metric based theory is diametrically self inconsistent from basic definitions. The most inexplicable anthropomorphic aspect of the old standard model was the merely dogmatic adherence to Einsteinian general relativity, known to be incorrect for almost a century, and conclusively shown to be so in UFT 101 to 199 in many different ways. This dogma was so tenacious that the dark matter theory was introduced to try to save the Einstein theory, even thought it has been known since the late fifties to be hopelessly unable to describe whirlpool galaxy velocity curves. It is now known that dark matter does not exist in nature.

In addition to the refutations of the metric based general relativity, UFT101–UFT199 gradually realized that the connection in geometry must be antisymmetric. The historical truth about the development of the geometry used by Einstein also became clear. It became clear that Riemann inferred only the concept of metric, and did so in the early nineteenth century. The concept of connection was developed by Christoffel in the eighteen sixties. Neither Riemann nor Christoffel inferred curvature, a concept due to Levi–Civita and others in the early twentieth century. Torsion was not inferred until the nineteen twenties, by Cartan and colleagues. So it became clear using accurate historical scholarship that the geometry used by Einstein is basically incorrect. It leaves out torsion. Einstein and others, notably Hilbert, developed general relativity in the decade 1905–1915, in an era when torsion was unknown. Unfortunately therefore their work cannot be correct – neither can a century of work in Einsteinian general relativity.

The final major advance made by ECE theory of unified physics occurred in UFT199, when a new equivalence principle was discovered and a new type of philosophical unification made of classical and relativistic physics. This advance has been given the appellation "frame dynamics", a completely self consistent general relativity developed from the well known fact that passive and active rotations are equivalent. Thus we return to Kepler's basic statement: ubi materia ibi geometria.

The overall progress made in UFT 101 to 199 is therefore clear, the ECE theory has advanced greatly in several directions, and has simplified and clarified. The progress made is summarized in this review in fifteen sections as follows, the acronym UFT denotes the unified field theory papers on www.aias.us. Section 2 reviews spin connection resonance and computer simulation of spacetime energy devices made in UFT107, UFT153 and papers by Eckardt and Lindstrom. Section 3 reviews the progress made in geometry and its applications in UFT101–104, 109, 112, 128, 137, 139, 142, 151, 152, 167, 185–189, and 199. Section 4 reviews progress made in astronomy and cosmology in UFT105, 108, 111, 120, 123–127, 148, 150, 151, 154 and 155. Section 5 reviews progress in precession theory in UFT110, 114, 117, 119, 145–147 and 149. Section 6 reviews the development of conservation, continuity and Poynting theorems from geometry in UFT 116, 121, 156, and 168–170. Section 7 reviews progress made in dynamics and fluid dynamics in UFT 113, 140, 143, and 149. Section 8 reviews the new equivalence principles in UFT 141 (derived from antisymmetry) and 199 (derived from passive and active rotations). Section 9 reviews the progress made in antisymmetry in UFT 131–134, and papers by Eckardt and Lindstrom. Section 10 reviews the progress made in ECE quantum mechanics, notably the fermion, quantum Hamilton and quantum force equations, and conclusive refutation of Copenhagen indeterminacy in UFT 129, 130, 135, 135, and 172–180. Section 11 reviews the refutation of standard particle scattering theory and its replacement by R theory in UFT 150, 155, 157–167, 171 and 180–182. Section 12 reviews the refutations of Einsteinian general relativity in UFT 150, 155, and 190–195. Section 13 reviews the new relativity of UFT 196–199. Section 14 reviews UFT 144 and 154 on the interaction of electromagnetism and gravitation and on electrodynamics. Section 15 reviews the application of B(3) theory to clean fuel technology in UFT 183 and 184. Section 16 finally reviews the various geometrical refutations of Einsteinian general relativity using the antisymmetry of the connection in UFT 122, 137 and 139.

2. Spacetime energy devices and spin connection resonance

UFT107 has received about ten thousand referrals to date, and is the most read paper on www.aias.us. A referral is the contemporary, real time, method of measuring interest in a paper as defined by Berners-Lee and co workers as is well known. In this paper experimental reports were considered theoretically of spacetime energy detected by a variable frequency Faraday disk generator. It was reported experimentally that the device exploded due to a surge of power. UFT107 explains this surge of power in terms of spacetime power and spin connection resonance. It uses a rotational tetrad and spin connection resonance equation which was solved numerically to give three different types of power resonance. The dynamics of the homopolar generator were studied and graphed. Notably, electron orbits were graphed for a variable frequency generator. Resonance curves were given for variable current, and a prototype discussed based on Tesla's work. An instrumentation diagram and analysis were included in the paper. This design promises to be the first commercial spacetime device, based on a modification of the well known Faraday disk or homopolar generator. UFT153 continued the study of spin connection resonance in a spherically symmetric spacetime using the now obsolete metrical method of general relativity in a spherical spacetime. UFT107 does not depend on the now obsolete metrical method of general relativity. Progress in the development of ECE theory has been rapid, the obsolescence of metrical general relativity was first realized clearly in the series of papers UFT 190 ff.

Traditional electromagnetic theory has not been able to effectively explain the numerous anecdotal reports of devices that on a local scale seem to produce more energy than they consume. The ECE electromagnetic and gravitational sub-theories offer a strong means of explaining some of these occurrences. Non-traditional effects have been predicted in several of the UFT papers to be very small, with one of the most recent (UFT144) predicting that the spin connection portion of the electric and magnetic fields is about 1/c time the traditional component, in SI units. Such an interaction would in all likelihood not be observable in normal electromagnetic phenomena. Resonant behaviour with this small component being the driving term has emerged as the dominant way of possibly explaining such events. Three types of resonant behaviour have been discussed:

1. Resonance where the full field equation set is used, or a subset of the full set where the rest were demonstrated or assumed to be inconsequential.
2. Resonant coupling of the electromagnetic field to the gravitational field
3. Resonance resulting from particle motions.

An example of the first is a classic example demonstrating the power of the ECE theory in explaining observations that defy rational explanations with traditional physics For example, from Paper 107 as discussed earlier, there have been reports of a Faraday disk generator exhibiting a powerful resonance effect hitherto unknown. Explanations were given by the ECE field equations expressed in terms of potentials. These are highly non-linear, hyperbolic, and coupled to each other, making analytic solutions impossible except for the simplest of examples. Numerical methods are inevitably involved with a variation on the Runge Kutta method used typically for one dimensional problems, and Finite Difference or Finite Element techniques being used if the problem requires two or more independent variables. Three types of possible resonances were identified. The first method was used to solve the two equation set for the Faraday disk. All solutions grew in time, even if the terms involving integration constants were set to zero. The second (in UFT107 the third) method utilized a zero crossing of the denominator of the current. The third type of resonance was discussed at length in papers 153, 167–170 where coupling of the electromagnetic field to the local gravitational field was made through the use of a local distortion in the metric. Paper 153, the forerunner to the development of the so called "metric method", looked at the use of the Hamiltonian and the Euler–Lagrange equations in a spherical space-time. We know now that this space-time is invalid, but this did demonstrate the emergence of the Euler–Bernoulli resonance as the mechanism in this form of field coupling.

The coupling the electromagnetic field to the gravitational field was developed in paper 167. The distortion of the metric from that of a Minkowski metric for a material-free space to that with material media being present shows up in terms of the relative permittivity and permeability properties of matter. Euler-Bernoulli resonance was demonstrated for a simple one dimensional problem. The gravitational Poynting theorem was introduced in Paper 168, and used in Paper 169 to demonstrate the energetic coupling between the electromagnetic and gravitational field. In Paper 170, the vacuum field as defined by the zero point energy of a harmonic oscillatory and the use of the Poynting theorem was used to explain the electric power absorbed along a straight wire from the vacuum. A discussion of current experiments and technology that utilize this mechanism was also included. A final paper [16] was presented to demonstrate the metric coupled resonance in an infinitely long solenoid subjected to a time dependent driving current. In this paper, a quadratic coupling of the magnetic field to the metric, as typified by a solenoid core undergoing dimensional change with magnetic field due to magnetostriction, or perhaps compression due to electromagnetic forces in the windings, is used to demonstrate resonant behaviour. The finite element method was used to solve the problem containing two independent variables. Heterodyning behaviour was also observed as the resonant point was approached.

Resonance in the motion of charged particles was discussed at length in papers 143, 144, 181–184. Following the development of fundamental dynamics from geometric principles in Papers 143 and 144, paper 181 introduced the concept of the ECE particle matter wave. This went on in Paper 182 where the Hamilton-Jacobi equation was used to describe the interaction between a particle and a field. Compton scattering of a photon from an electron was discussed showing that during the interaction, the photon acquires a finite interacting mass. A correction to the deBroglie–Einstein equations was also given.

Paper 183 was a discussion on molecular dissociation due to the B(3) field. This longitudinal magnetic field was shown to produce a driving torque by interaction with a molecule or ion. This torque could be amplified in an Euler type resonance to induce molecular dissociation. Application to industrial processes were discussed culminating in a separate paper UFT184 devoted to a specific (Kurata) process which for example is used to breakdown complex hydrocarbon wastes into useable diesel oil that burns with less emissions than commercial diesel generated from crude oil sources.

In a series of four papers it was shown that for free space conditions using the single polarization ECE engineering equations, that the ECE equations of electromagnetism are equivalent mathematically to the Maxwell–Heaviside (MH) equations at the potential level. This of course is not possible on a physical level, since the MH theory presupposed a flat, torsion free space-time, totally incompatible with ECE theory. In the first paper to discuss this reduction, [12] it was shown that in the static case, for a single polarization, the ECE theory reduced mathematically to the MH theory. This was expanded [13] to include dynamic situations where the vector spin connection was parallel to the vector potential. In the second paper in this series [14] it was shown that this reduction occurred whenever the scalar potential was a separable function of space and time. In the third paper [15] it was shown that if a non-ECE solution existed, but at some point even if only momentarily, the solution became Maxwellian and then it remained Maxwellian. It was speculated that the only possibility for non-Maxwellian behaviour was in the instance of discontinuous potentials, that is, potentials that are not-separable and cannot be represented by Fourier series. It should be pointed out that this reduction is only applicable to free space conditions. The introduction of material properties, such as magnetization, especially if such properties depend upon the fields themselves, then this reduction is not valid, as was demonstrated in the metric papers on electromagnetic theory (UFT167–UFT170).

3. Geometry

Two or three years ago when papers UFT101 onwards were being written, it was not known that the connection was antisymmetric in its lower two indices and the complete incorrectness and resulting obsolescence of the metric based Einstein general relativity was not known. Some of the papers reviewed in this section dealt with the detailed proofs of the Cartan identity, from which the field equations of ECE theory are obtained. Methods were also developed to simplify the poofs of the Cartan structure equations. This type of paper started with UFT88 and UFT99, papers in which the proof is written out in full detail, and gradually became simpler and stronger in papers such as UFT137, UFT139 and UFT167. The original identity given by Cartan related the torsion and curvature, but was given by him in an elegant but concise format, very difficult for non-specialists to understand. Very few textbooks of the twentieth century general relativity even mentioned torsion and it gradually became clear during the course of development of ECE theory that the structure of Cartan geometry was unknown to these twentieth century authors. They considered Riemann geometry as a subject in which only the curvature was defined. Some of them did not appear to know that Riemann inferred only the metric and referred to the connection and curvature as being due to Riemann. In historical fact it was Christoffel who inferred the connection in the eighteen sixties, and Levi–Civita and contemporaries who inferred the idea of curvature in the early years of the twentieth century. It was not until the nineteen twenties that torsion was defined by Cartan and co workers. So it took almost a hundred years to progress from the ideas of Riemann to those of Cartan. It was the latter who finally inferred the correct description of the general mathematical space in terms of both torsion and curvature. It is incorrect to omit torsion and keep curvature, or vice versa.

It also became clear gradually that the twentieth century idea of general relativity was based on the axiomatic but incorrect use of a symmetric connection, by which is meant a connection that is symmetric in its lower two indices. Some textbooks started with this axiom and did not mention torsion at all. The symmetric connection was often referred to as the Riemann connection, the authors did not seem to know that the idea of connection was not known to Riemann at all, and that he never used it. The only book that defined torsion and curvature correctly and clearly, in a manner that was not plagued by nearly incomprehensible mathematical abstraction, was that of S. P. Carroll [11]. The torsion and curvature tensors are always defined by the commutator of covariant derivatives acting on a vector, or more generally, any tensor. This definition is true in any mathematical space in any number of dimensions. This procedure is equivalent to the two structure equations of Cartan and Maurer, which defined two differential forms of geometry known as the torsion and curvature forms. The structure equations are again elegant but so concise as to be incomprehensible to anyone except the ultra specialist. Before the development of ECE theory they were almost unknown in physics.

In order for these important structure equations to become comprehensible to non-specialists they had to be developed during the course of ECE theory into tensor equations, and then into vector equations, whose structure was gradually simplified for use in practical engineering, notably spin connection resonance. It also had to be shown that they were rigorously equivalent to the action of the commutator in always generating the torsion and curvature. This equivalence is demonstrated through the very obscurely named tetrad postulate. In fact this is a very fundamental result of the invariance of the complete vector field, and not a postulate. So rigorous proofs of the tetrad postulate also had to be given, and Cartan's concise notation explained clearly. These proofs gradually evolved, clarified and simplified in the papers reviewed in this section, and listed in the introduction of this review paper. The two structure equations define curvature and torsion through the mediation of the spin connection of Cartan, a development of the original idea of connection due to Christoffel in the eighteen sixties. Neither Riemann nor Christoffel inferred the idea of curvature, they did not use tensors, which were inferred in about 1900 by Levi–Civita, Ricci, Bianchi and contemporaries. The first inference of the idea of spacetime curvature seems to be due to Levi–Civita in the early years of the twentieth century using the then new tensor methods. This inference did not use, or at least did not clearly define, the idea of the commutator of covariant derivatives, the curvature was defined in isolation. The commutator encapsulates the idea of a round trip around a closed loop in a mathematical space. There are two round trips considered, clockwise and anticlockwise.

It is very important to realize that the round trips in different senses produce a change in the orientation of a vector. The usual example given is the surface of a sphere. If a vector is parallel transported to the north pole along a line of longitude, it arrives with a given orientation. Parallel transport in this context means transport in such a way that the vector is tangent to the line of longitude at each point of transport. If the procedure is repeated by transporting first along the equator and then along another line of longitude to the pole, it arrives with a different orientation. The vector has been operated upon and the operator depends on the paths taken in a given mathematical space. If the space considered is a circle defined in a plane, transport clockwise and anticlockwise has no effect on the vector. So in a flat space there is no effect. The difference between a flat and general space is that in the former, the connection is zero, in the latter is non zero. The connection is used to define a derivative, one which retains its format under all possible coordinate transformations. This is known as the covariant derivative, and it is the sum of the ordinary derivative and a spin connection term. The difference between two round trips, clockwise and anticlockwise, is represented mathematically by the commutator of two covariant derivatives. The commutator is an operator and acts on the vector being transported.

The result of this operation contains a combination of connections, ordinary derivatives of connections, and connection products. Combinations of these terms are used to define the torsion tensor and the curvature tensor and all the details of this procedure were given in UFT99. These details were almost never given in twentieth century textbooks in general relativity, and only partially by Carroll [11]. A significant feature of the development of ECE theory is that proofs are given in all detail, and that these proofs are simplified and improved. The torsion tensor is a linear combination of two connections with lower indices reversed and this definition is equivalent to the one given in the first Cartan structure equation. The lower two indices of each connection are the same as the indices of the commutator. The latter is by definition antisymmetric, meaning that it changes sign if the indices are interchanged. So the connection is also antisymmetric in its lower two indices. This was first realized in UFT122, more than 150 years since Christoffel inferred the connection in the eighteen sixties. Following upon this realization some more proofs were developed of the antisymmetry.

Why did it take 150 years to realize this? The answer to this question is that when Christoffel inferred the connection the idea of a round trip was not available, the idea of curvature was not yet inferred. It would take a further forty years for curvature to be inferred. In the eighteen sixties it was known only that the metric of Riemann (1826–1866) was symmetric, and then only to a very few ultra specialists in mathematics. There was no way of knowing the symmetry of the connection in its lower two indices, or even of knowing that it has any particular symmetry. It must have been assumed that it was symmetric, and assumed without proof. This was a bad error because any matrix is in general asymmetric, not symmetric. Therefore when the idea of curvature arrived in the early twentieth century, the connection was assumed by bad habit (i.e. dogma) to be symmetric. This basic mathematical error was repeated endlessly, even after Cartan had introduced his structure equations, and even after they were pointed out to Einstein by Cartan. Unfortunately this dogma means that the entire twentieth century era in general relativity must be discarded as being incorrect. This dogma means that the torsion was incorrectly discarded, because a symmetric connection means that torsion is incorrectly zero. Zero torsion means zero commutator, and no curvature either, and zero torsion and curvature mean a flat spacetime. So any symmetric connection is zero, because a flat spacetime has no connection. The commutator of ordinary derivatives in a flat spacetime is zero and a flat spacetime has no torsion and no curvature.

The negative influence of dogma continued to linger in the development of ECE theory up to UFT122 when it was finally realized that the symmetric connection vanishes in all mathematical spaces. This is one of the major mathematical advances of ECE theory. The papers up to UFT122 are rigorously correct, but UFT122 is the one that discards the symmetric connection and points out that the connection wherever it occurs is always antisymmetric. Papers up to UFT122 make no statement about the symmetry of the connection other than to assume implicitly that it is asymmetric (i.e. the torsion is assumed to exist). Even that though was a great improvement over the dogma that used without reason a symmetric connection with no torsion.

Once it was realized that the connection is antisymmetric, other inferences rapidly followed and developed into the antisymmetry laws of ECE theory, laws that directly refute the standard model both of electrodynamics and gravitation. The antisymmetry laws are reviewed in another section of this paper and have been greatly developed by Eckardt and Lindstrom. UFT186 to UFT189 used the antisymmetric connection with metric compatibility to develop a general relativity based exclusively on torsion. In UFT190 and UFT192 to UFT195, however, the development changed direction in another fundamentally important way, devising direct and simple means of refuting metric based general relativity completely, so that the Einsteinian general relativity disintegrated. For example it was shown that the force law of a precessing ellipse is not that used in Einsteinian general relativity; it was shown that the falsely attributed "Schwarzschild metric" does not produce a precessing ellipse, and finally in UFT194 it was shown in a very simple way that Einsteinian general relativity is fundamentally self inconsistent in all spherical spacetimes. In earlier papers such as UFT150 and UFT155 it had been shown that Einstein's calculations of light bending and Shapiro's calculations of time delay are erroneous. Historical scholarship during the course of development of ECE theory had shown that Einstein's general relativity has been refuted by many of the best scientists, notably Schroedinger, Eddington, Dirac, Levi Civita and others. Finally this type of scholarship showed that Schwarzschild had refuted Einstein's perihelion calculation as early as December 1915. It is also obvious why the dogmatists ignored the science, to admit to their theory being dogma would have been a monumental loss of face and funding. However, professional scientists have calmly accepted the end of the Einsteinian era, that much is very clear from the feedback methods devised by Berners-Lee et al.

In UFT199 another major advance in the geometry has just been made and given the appellation "frame dynamics". In retrospect it is clear that any type of motion of a particle with a fixed frame and no connection is equivalent to the motion of the frame with fixed particle and no ordinary derivative present, the motion of the frame (frame dynamics) being described entirely by the connection: Kepler's "Ubi materia ibi geometria". A very well known example is that of the equivalence of the active and passive rotation of a vector in a plane. UFT199 promises to result in an entirely new and rigorously correct general relativity based on rationalizing observation with geometry in the manner of Kepler. This development is particularly timely because metric based general relativity can no longer be used. It is incorrect for all spherical spacetimes. So ECE papers that utilized metric based general relativity will have to be redeveloped using frame dynamics. The only metrics that remain valid are the Minkowski metric of special relativity and the Crothers metric.

4. Astronomy and cosmology

In the UFT papers 105, 108, 111, 120, 123–127, 148, 150, 151, 154 and 155, several developments were made of astronomy and cosmology. In this phase of development of the ECE theory of unified physics the complete disintegration of the metrical general relativity in UFT194 was obviously not known. The complete incorrectness of the metrical method for all spherical spacetimes became known in UFT194, and attempts to replace it began in UFT196 ff. Such is the rapidity of development of ECE theory that all this has occurred within the space of two or three years, and is a major paradigm shift in physics. The metric based papers in this section are UFT 105, 108, 111, 148, 150, 151 and 154. They refuted the Einsteinian theory by use of torsion, incorporated in various contexts into cosmology. In view of UFT194, these papers and methods will be revised in future work based on a completely new approach, one which uses none of the ideas of the obsolete standard model, and none of the ideas of Einsteinian general relativity. At the time of writing (November / December 2011), it seems that the frame dynamics developed in UFT199 is the best way forward, because it is a fully relativistic theory based on a new and very powerful equivalence principle.

The obsolete Einsteinian theory was based on the use of a metric which was incorrectly, or falsely, attributed to Schwarzschild. It was commonly known as "the Schwarzschild metric". In UFT190 ff, which emerged in late 2011, it was shown to be complete nonsense. So the metric based papers 105, 108, 111, 148, 151 and 154 must now be regarded at the time of writing (November / December 2011) as transitional work which was intended to improve the then known metrical methods (methods that omitted torsion). For example UFT105 attempted to produce a new theory of light bending, UFT108 produced a new theory of binary pulsars without use of gravitational radiation, UFT111 and UFT148 attempted to produce new orbital theorems, and so forth. UFT120 was a comprehensive refutation of black hole theory using torsion, and is the one which has produced the secondmost number of referrals to date, several thousand. UFT120, 123–127, 150 and 155 are not affected by the complete collapse of metrical general relativity that emerged in UFT190 ff. UFT123–UFT127 produced a theory of whirlpool galaxies and so on based on spacetime torsion, and should be regarded as steps towards the understanding that is now beginning to emerge in UFT199. UFT150 refuted the Einstein theory of light deflection by gravitation in several ways, and essays and broadcasts based on it have become very well known, notably the essay "Nobody's Perfect" by co author Horst Eckardt, broadcast by Robert Cheshire and myself. Finally in this section UFT155 refuted the Shapiro theory of time delay. Following upon UFT194 the theory of gravitational light deflection and time delay is now known to be completely incorrect at a fundamental level. So UFT150 and UFT155 are early signs of the final refutations of UFT190 ff.

5. Precession theory

Precession theory was developed in UFT110, 114, 117, 119, 145–147 and 149 using mainly metrical theory but also field theory. Following upon UFT194, the only valid metrical theory of general relativity is that based on the Minkowski and Crothers metrics. UFT110 developed Thomas precession in terms of the Minkowski metric, and UFT149 developed equations of motion in terms of the same metric. These two papers are still valid without further development. UFT114 developed a theory of the gravitational red shift but it was based on a metrical theory that is now obsolete following upon UFT194 and the gravitational red shift can only be addressed now by a development of a theory such as that of UFT199, now in its earliest phase. Tens of thousands of papers and monographs on the Einsteinian general relativity have also been made obsolete, of course, by UFT194 and UFT190, 192, and 193, and ECE theory of unified physics in general. This is why the science being developed at the time of writing is a major paradigm shift. UFT117 developed a theory of the earth's gravitomagnetic precession. The parts of UFT117 that depend on the ECE field equations are rigorously valid, but the parts that depend on the twentieth century idea of metric based general relativity have been invalidated by UFT194 and must be redeveloped in future work. The theory that Gravity Probe B set out to "confirm" has been shown to be completely incorrect by UFT190, and 192 –194. This was the theory based on the falsely attributed Schwarzschild metric. Everything based on that metric is incorrect, and also everything based on any spherical spacetime metrical method of the now obsolete general relativity. This is why the conclusions of UFT194 represent a major paradigm shift that has only just emerged.

UFT119 set out to developed a theory of the equinoctial precession and of galactic dynamics, and again the parts of that paper that depend on the ECE field equations are correct, while those parts that used the now obsolete metrical method are obsolete because of UFT194. All these papers are correct mathematically, but UFT194 showed up a fatal self inconsistency in the very foundations of metrical general relativity. This fatal self inconsistency was not known until late 2011. One of the major implications of UFT194 is that satellites sent up to test an obsolete theory will produce data that should be used to test a new, rigorously correct, general relativity free of any Einsteinian concept. Experimental data from these satellites are always useful, but it is pointless to go on claiming that they verify an incorrect theory, Einstein's general relativity. UFT145 to 147 dealt with the Sagnac interferometer. The parts of these papers that are based on the Minkowski metric can be used without further development, but not those parts that use a metric that is different from the Minkowski metric, because UFT194 has shown that all such metrics are self inconsistent. Earlier developments of the Sagnac effect in UFT1 to UFT99 are still valid, because they rested on the new ECE field equations, and not on the twentieth century metrical method.

Another major discovery of ECE theory is that any metric inferred from the Einstein field equation is incorrect because of the omission of torsion, and this is summarized in the monograph [1 –10], "Criticisms of the Einstein Field Equation" published in early 2011.

6. Conservation, continuity and Poynting theorems

These are developed in UFT116, 121, 156, and 168 –170. In UFT116 the ECE equations of dynamics are developed in vector notation in parallel with the equations of electrodynamics. The continuity equation is developed from the tetrad postulate of Cartan geometry as a special case. In UFT121 the Noether theorem is deduced from the tetrad postulate, which is therefore made the basis both of energy conservation and continuity in physics. UFT156 was written as an attempt to develop a metric based theory of electrodynamics, but as with all metric based theories it is refuted by UFT194. A Poynting theorem for gravitation was developed in UFT168 in parallel with the idea of a Poynting theorem for electromagnetism. These Poynting theorems are therefore geometrically based and depend on the first structure equation and Cartan and Evans identities. In UFT169 a method was developed to describe the deflection of light by gravitation with the Poynting theorem, and this seems to be the only valid method of doing so in view of UFT194. Finally UFT170 developed a theory of the vacuum with the Poynting method.

7. Dynamics and fluid dynamics

These are papers of field equation theory unaffected by UFT194 because they do not use metrical theory. They are UFT113, 140, 143 and 149. UFT113 developed a set of dynamical equations from the Cartan identity and the Evans identity, proven in UFT137, UFT139 and UFT167. UFT140 applied this method to concepts of fluid dynamics, and UFT143 developed new concepts in fundamental classical dynamics by incorporating terms from the spin connection. The geometrical concepts used in these papers are the first Cartan structure equation, the Cartan identity and the Evans identity. Hypotheses are made to produce dynamics from the geometry, and as usual, the hypotheses must be tested with experimental data. The underlying geometry of Cartan is unchanged, the Evans identity being an example of the Cartan identity itself. In the Evans identity, Hodge duals of tensors are used together with the Hodge dual of the antisymmetric connection. In four dimensions the Hodge dual of an antisymmetric tensor is another antisymmetric tensor [1–10]. The Hodge dual of the antisymmetric connection is defined as in UFT167 and earlier papers by the antisymmetric indices of the connection. The latter is not a tensor, but its Hodge dual can be defined in this way, and the Hodge dual produces another well defined antisymmetric connection. The importance of these new mathematical methods is that the Cartan identity gives the inhomogeneous field equations both of dynamics and fluid dynamics, as well as those of electrodynamics. These structures are summarized in vector format throughout the development of ECE theory, notably in the engineering model. The Evans identity gives the homogeneous field equations and their structures. The field equations rest rigorously therefore on geometry that correctly includes torsion. Cartan geometry seems to be adequate for physics at the time of writing. Other geometries may be used provided they account both for torsion and curvature.

8. Equivalence principles

Two types of equivalence principle have been inferred, in UFT141 and UFT199. UFT141 rested on the first Cartan structure equation and used antisymmetry to infer the equivalence principle assumed by Newton – the equivalence of gravitational and inertial mass was therefore proven mathematically for the first time. Antisymmetry is inferred from the antisymmetry of the Christoffel connection proven as in Section 3 of this review paper. The equivalence of inertial and gravitational mass has been proven experimentally to phenomenal accuracy, so UFT141 represents a very accurate experimental proof of the antisymmetry first inferred in UFT122. It also proves that there is no such thing as a symmetric Christoffel connection. Unfortunately the whole of twentieth century general relativity was based on a symmetric connection and is therefore meaningless. From a continuous study of feedback, which shows how ECE theory is being received, this astonishing paradigm shift has been accepted, essentially throughout the entire university sector worldwide.

UFT199 has just been completed and introduces a completely new equivalence principle based on the equivalence of active and passive rotations in a plane. The rotation of a vector clockwise in a static reference frame is equivalent to the anticlockwise rotation of the frame itself, keeping the vector static. In UFT199 this well known fact of simple geometry is generalized into Cartan geometry, considering in the first instance the definition of the covariant derivative in Cartan geometry. It is inferred that the ordinary derivative of a vector is equal to the spin connection acting on the vector, these being the two terms of the covariant derivative. The ordinary derivative generalizes the rotation of the vector in a static frame, the spin connection generalizes the idea of a rotating frame with static vector. This equivalence principle leads to a rigorous and completely new relativity known as "frame dynamics". In UFT199 an example is given in which the solar system orbit is described with spin connection components, so the orbit is understood as the motion of the frame –frame dynamics, and the orbit becomes pure geometry –"ubi materia ibi geometria" of Johannes Kepler.

Frame dynamics is capable of great development and is the result of another major change of thinking (or paradigm shift) in physics –orbits can be predicted neither by Newtonian theory nor by the incorrect Einstein theory. In frame dynamics the observation of orbits in astronomy is rationalized in terms of geometry as described in Section 3 of this review paper. This is exactly what Kepler did with the orbit of Mars, he showed it to be an ellipse. It is well known that the Newtonian dynamics fails to provide the centrifugal force. This was shown conclusively and clearly, and astonishingly for the first time, in Section 3 of UFT196 by starting with the functional equation of an ellipse and deriving the force law in cylindrical polar coordinates: a very simple method but one that gave a fundamental new insight. The force law was found to be negative valued, directed along the radius unit vector, and inversely proportional to r squared. It was an inverse square law, but purely attractive, so that an object of mass m would fall in to an object of mass M in the old thinking. The latter is a disaster because the orbit would collapse. In frame dynamics the orbit is due to the ever present motion of the frame itself. There is no anthropomorphic "attraction" between m and M, there is no universal gravitation. If the ellipse is made to precess the force law is different, it becomes the sum of inverse square and inverse cube terms.

9. Antisymmetry

The first papers on antisymmetry are UFT131–134, and were applied to electromagnetism and gravitation. These papers developed from the important inference in UFT122 that the Christoffel connection is antisymmetric, another major paradigm shift in physics that has been accepted by the university and other sectors worldwide. The tetrad postulate of Cartan geometry means that the Christoffel connection can be developed as the sum of two terms, and this fact of geometry led to the antisymmetry laws of ECE theory and the Lindstrom constraint theory. This part of ECE theory has been developed extensively by Eckardt and Lindstrom and applied to practical engineering problems. These results by Eckardt and Lindstrom are reviewed in this section. The first major result of the antisymmetry laws was the refutation of gauge theory, which had already been abandoned in ECE theory in papers following UFT71. The obsolete U(1) gauge theory of the standard model was based on the diametrically self contradictory "massless photon" and the idea of the invariance of physical properties under a U(1) gauge transformation. Unfortunately that was a theory without a spin connection, a theory developed in the flat Minkowski spacetime. As soon as the spin connection is introduced there is no gauge invariance. Without the spin connection there is no generally valid relativity. Twentieth century physics was a dogmatic construction based on an incorrect general relativity for gravitation, and an incorrect gauge theory of electromagnetism, a gauge theory that was not only incorrect, but one of special relativity.

As soon as antisymmetry was applied rigorously to the old field tensor of electrodynamics, it split into two components, one being the negative of the other. The consequences of this result were a disaster for U(1) gauge theory, in consequence of its dogma the Maxwell Heaviside theory itself became meaningless as shown in UFT131–134. It was also shown that the only way out was to adopt Cartan geometry and the ECE hypothesis introduced in the first papers of 2003 that transformed Cartan geometry into a generally covariant electrodynamics.

Antisymmetry proved to be a useful concept for engineering.

The biggest single advantage offered by the antisymmetry equations (UFT132,133,134) when applied to practical application of the ECE equations for electromagnetism was the addition of six new equations to the equation set, providing enough additional equations to solve field problems without making drastic assumptions regarding the form of the solutions. The electric antisymmetry equations provided two equivalent definitions for the electric field which in essence removed two variables from the equation system. The magnetic antisymmetry equation provided a constraint on the spin connection and vector potential. This was expressed in a differentio-algebraic relationship by Lichtenberg (to be published in Journal of Foundations of Physics and Chemistry).

The Lindstrom Constraint was introduced in UFT133, 134 as a sub-set of the magnetic antisymmetry equations in an attempt to reduce the number of variables paralleling what was experienced for the electric antisymmetry counterpart. This constraint proved to be too limiting however since it limited the magnetic vector potential to that of the ECE vacuum state, and so was later retracted for single polarization electromagnetics [13]. This constraint will be re-examined during the development of electromagnetics using multi-polarization techniques in future papers.

The existence of orbital and spin forms for the torsion tensor was introduced in Paper 134. This was developed into an electromagnetic theory in Paper 144.

The ECE vacuum state was defined and explored in two papers [18,19]. The vacuum state is what remains when the ECE electric and magnetic fields are set to zero. This leads to solvable equations only if all antisymmetry laws are included. The vacuum state was shown to consist of a potentially energetic family of plane waves for the vector and scalar potentials. One family of solutions that consistently emerged was in the form of a hyperbolic tangent. This offered propagating step functions for the vector potential. The hyperbolic tangent form of solution was also observed for the ECE component of Coulomb's Law which had the form of a topological charge density. This offered a hint at the possibility of a virtual particle model, even though the ECE theory for electromagnetism is a field theory.

The second vacuum paper [19] demonstrated that the existence of the vacuum state does not normally influence the electromagnetic state of a device. The electromagnetic state of a device for example, simply floats upon the vacuum state. It was also demonstrated that the speed of propagation of the waves in the vacuum state was light speed for empty space. It was suggested that energetic interaction of a circuit with the vacuum could only occur if a change in geometry (boundary conditions) for the circuit occurred.

10. The unification of quantum mechanics and general relativity, refutation of Copenhagen indeterminacy and the Dirac equation

These advances were made in UFT129, 130, 135, 136 and 172–180. Prior to these advances, unification of quantum mechanics and general relativity was achieved in the first ECE papers in 2003, when it was shown that the wave equations of physics emerge from the tetrad postulate of geometry. The Dirac equation was simplified in UFT129 and 130 into an equation with 2 x 2 matrices rather than the 4 x 4 Dirac matrices. These 2 x 2 matrices were given a geometrical interpretation within the context of Cartan geometry. UFT129 developed the rest fermion, and UFT130 the fermion in motion. UFT135 and 136 extended this analysis to SU(2) electrodynamics and SU(3) nuclear theory. At that stage of development the Dirac equation itself was not criticised, indeed the Dirac equation had been derived earlier in UFT4 but the meaning of its negative energy interpretation questioned.

In UFT172 the fermion equation was developed to show that it has rigorously non-negative eigenvalues, thus eliminating from physics the obscurities that result from the Dirac sea interpretation, providing for the first time a single particle fermion equation. The half integral spin, Lande factor, Thomas factor, Darwin term, ESR, NMR, fine and hyperfine structure and identically positive probability current were deduced from the single particle fermion equation. Much of the self destructive obscurity of twentieth century physics resulted from the need for a quantum field theory that was a multi particle theory. The latter in turn arose from the use of the original Dirac matrices. These produced what were apparently negative energy eigenvalues. In contrast the fermion equation produces rigorously positive eigenvalues and refutes the use of the Dirac matrices for this reason. In future work quantum field theory can be developed as a single particle theory that is simpler and free of the dogma of negative energy. In UFT173 the fermion equation was given in a simple covariant format and the half operator method of solution introduced. In UFT174 solutions of the fermion equation were given for the hydrogen and helium atoms, and the Pauli exclusion principle deduced straightforwardly from the fermion equation using parity inversion.

In UFT175 the quantum Hamilton equations were inferred in which position and momentum are defined simultaneously, thus refuting the Copenhagen indeterminacy at the outset. The untenable nature of indeterminacy was demonstrated with a novel anti-commutator method. In UFT176 higher order commutators were used to further refute the Copenhagen indeterminacy in a particularly clear and simple way. In UFT177 this work developed into the inference of a novel force equation of quantum mechanics, and force eigenvalues illustrated. UFT178 gives a relativistic quantum force equation and developed spin orbit coupling. UFT179 gives a generally covariant fermion equation, and UFT180 develops the idea of a wave equation derived from the tetrad postulate.

These papers represent another major paradigm shift in quantum physics, eliminating the dogmatic twentieth century adherence to negative energy. "Negative energy", a contradiction in thought, was shown to result from the wrong choice of matrices by Dirac in his first papers. The correct choice of matrices follows from the fermion equation.

11. Photon mass and refutation of Standar Model particle scattering theory

These major advances and refutations are given in UFT150, 155, 157 –167, 171, and 180 –182. The original motivation for these papers on particle scattering theory was an attempt to devise new experiments to measure the mass of the photon. The first attempt was made in UFT150, which was originally intended as a development of Einstein's calculation of light deflection due to gravitation to include finite photon mass. However, it was found that Einstein's calculation contained errors, self inconsistencies and obscurities which are summarized in UFT150 and in the popular broadcast "Nobody's Perfect". Following upon UFT194, it is now known that the Einstein method is fundamentally self inconsistent and meaningless, because it is a metric based method that uses the incorrectly attributed and meaningless "Schwarzschild metric". The outcome is that gravitational light bending must be looked at again with an entirely new relativity, such as that of UFT199. The experiment and calculation that catalysed Einstein into fame are meaningless dogma. This conclusion is sometimes referred to as the post Einsteinian paradigm shift. In UFT155 it was shown that the Shapiro time delay calculation is erroneous and that there is no self inconsistency between light bending and time delay. Following upon UFT194 it is now known that the time delay method is meaningless because it is a self contradictory metric based method. It is not known why dogmatists continued to refer to these experiments as "precision tests" of a theory when that theory is so riddled with errors known since December 1915, errors pointed out repeatedly by leading scientists and culminating in the systematic, detailed and conclusive refutations of ECE theory reviewed in this paper.

In UFT158 the usual theory of Compton scattering of a massless photon from an electron was re examined with a view towards the investigation of photon mass. The correct equations were given of the scattering of a massive photon from a massive electron, using the fundamental postulates that attempted to unify special relativity and quantum mechanics, the de Broglie Einstein postulates of about 1922, based on the doctoral thesis of de Broglie. By giving the correct, incisively deduced, equations for the first time, the theory was found to be severely self inconsistent and was found to give a mirage of correctness when the photon mass is set to zero. The award of a Nobel Prize to Compton was based on this mirage. Not for the first or last time. In UFT159 and 160 the general theory of Compton scattering was developed and became wildly incorrect when tested with sufficient rigour. The first attempts were made in these papers to replace the theory with one based on the ECE wave equation, given the appellation "R Theory". In UFT161 the original two de Broglie postulates were augmented with a third, which made the mass of a particle proportional to the parameter R of the ECE wave equation. These were given the appellation "October Postulates" of October 2010. In UFT162 these incisive criticisms were extended to the Einstein absorption theory, which again relies on the idea of massless photon. As soon as photon mass is introduced the Einstein theory of absorption becomes completely untenable because it did not consider conservation of momentum. So another aspect of standard physics was easily refuted completely. A similar outcome occurred for the theory of Raman scattering. This situation was remedied in UFT`163 to 166 with the development of R theory and R spectra for atoms and molecules. A lot of further development is possible along these lines.

In UFT171 the electron positron collision theory behind such experiments as LEP and CERN was shown to be wildly incorrect and a self consistent R theory given. The basics of the particle scattering experiments made at such huge expense are incorrect. Their fine details cannot be correct, and as yet there is no sign of a Higgs boson. This basically incorrect theory has been very heavily parameterized (with as many as about 25 adjustable parameters), and obscure concepts such as symmetry breaking added to give a meaningless, tasteless soup. With so many parameters anything can be fitted, and as shown in UFT171 the basic theory is wildly wrong. Only R theory can be used to test data of any interest from particle colliders.

UFT167 was an attempt to develop a metric based theory of electrodynamics, but UFT194 means that UFT167 must now be regarded as a transitional paper, for the same reason as any metric based theory with the exception of one based on the Minkowski and Crothers metrics. Finally UFT180 –182 develop R theory in several directions.

12. Systematic refutations of Einsteinian general relativity

These systematic, simple and powerful refutations should end the era of Einstein's general relativity. The main ones are UFT150 and 155, and UFT190 and 192–195. The refutations in UFT150 and 155 are confirmed conclusively by UFT194, because the Einstein theory of light bending and the Shapiro theory of time delay are metric based theories. All such theories have been refuted in UFT194 in all spherical spacetimes. In addition the mathematical details of UFT150 and UFT155 are correct in all respects and have been accepted as such by the entire profession. In UFT190 the analytical function of the precessing ellipse was compared with the result from the Einstein theory and several inconsistencies found within the Einstein theory. At this point it was realized that the Einstein theory does not produce a precessing elliptical orbit. This point was made by Schwarzschild in a letter to Einstein of December 1915, Schwarzschild showed that Einstein's theory of perihelion precession was incorrect only a month after Einstein published it. In UFT192 it was shown conclusively that the function incorrectly attributed to Schwarzschild cannot produce a precessing ellipse. The function needed to do that within a metrical theory is much more complicated. So it became obvious that the Einstein theory is deeply flawed. In UFT193 it was shown that the force law of lagrangian dynamics needed to give a precessing ellipse is not that claimed in the Einstein theory. It is important to note that in

UFT193, the same lagrangian methods were used as those in the Einstein theory. So there is no way out for the latter. The true force law for a precessing ellipse is a sum of inverse square and cube terms, the one claimed in the Einstein theory is a sum of inverse square and inverse fourth order terms. This is a clear, simple and conclusive refutation of the Einstein theory. This refutation was confirmed by use of the computer to demonstrate the converse refutation conclusively: that a force law consisting of a sum of inverse square and fourth order terms does not give a precessing ellipse at all. It gives a very complicated function that is totally different from a precessing ellipse. This is conclusive confirmation of Schwarzschild's refutation of Einstein in December 1915.

At this point it became clear that there is a fundamental flaw in the very basics of the metrical method used to construct the line element in Einstein's relativity. In the important and milestone paper UFT194, this flaw was found in an astoundingly simple way by taking the definitions of the total energy and angular momentum obtained by the Lagrangian methods of the Einstein theory. These definitions were used to find an expression for the function which had been falsely attributed to Schwarzschild. The false function is one of r, the radial coordinate of the cylindrical polar system in a plane. In UFT194 it was found that this cannot be true, the relevant metrical function is a constant for all spherical spacetimes, thus refuting all of Einsteinian general relativity directly and in a conclusive manner. Why has it taken nearly a century to find this fatal flaw? The answer is that Einstein general relativity became an example of Langmuir's pathological science, an example of repeated dogma, repetition made carelessly without scholarship, thus heavily damaging the progress of science. In UFT195 it was shown that only the Crothers metric of a very general spacetime survives the refutation of UFT194.

14. Towards a new general relativity

Attempts were made to search for a new relativity in UFT196–UFT199. In UFT196 the methods of special relativity were used to construct a relativistic kinetic energy which was used in a lagrangian method to describe the observed orbit of a planet, a precessing ellipse. In Section 3 of UFT196 a simple refutation of the Newtonian description of orbits was found simply by differentiating the function of an ellipse as described already in this review paper. This method showed that the centrifugal force, so called, does not exist, and so the Newtonian description is refuted directly. In UFT197 the rotational Hooke law was implemented in a new approach to general relativity, and in UFT198 the new relativity was used to give a simple explanation of the velocity curve of a whirlpool galaxy. In UFT199 a new and powerful equivalence theorem has been forged which allows a fully relativistic treatment of all orbits, UFT199 has been reviewed already in this paper.

15. Interaction of fields and electrodynamics

Two papers of the series 101 to 199 have been dedicated to the subject of interaction of fields and electrodynamics, UFT144 and UFT154. UFT144 developed orbital and spin electric and magnetic fields in classical electrodynamics, a development based on the first Cartan structure equation and the Cartan and Evans identities. There may be some experimental evidence for the existence of such fields. UFT154 was based on a metrical method refuted in UFT194, and must be regarded as a transitional paper.

16. The development of Kurata / B(3) technology

Recently the fundamental B(3) field has been utilized in a number of new technologies that produce clean burning fuel, clean water from polluted sources, and fuel from sea water mixed with residuals. The first industrial scale plant based on Kurata / B(3) technology has been opened in Cordoba in Spain, initially producing forty thousand tons of clean burning diesel a year. So the B(3) field is used routinely in industry. UFT183 and 184 were written to support this new technology by providing an explanation of how B(3) helps the process of converting waste products into clean burning fuel. There are plans for rapid growth of this clean fuel technology, especially the ability to provide new energy from sea water mixed with residuals.

17. Geometrical refutations of the Einstein general relativity

As discussed already in this paper, the existence of torsion and the antisymmetric connection is enough to refute the Einstein theory directly. The most relevant papers in this context are UFT122, 137 and 139. The whole series of two hundred ECE papers to date can be viewed as numerous refutations of the Einstein theory, but more importantly, as the emergence of the first unified field theory in the history of physics. In view of the overwhelming acceptance of the theory, it is known now as the ECE theory of unified physics.

References

[1] M .W. Evans, Ed., J. Found. Phys. Chem. 2011 onwards, six issues a year, Cambridge International Science Publishing (CISP) (www.cisp-publishing.com).

[2] M .W. Evans, H. Eckardt and D. W. Lindstrom, "Generally Covariant Unified Field Theory" (Abramis Academic, 2005 to 2011), in seven volumes.

[3] M .W. Evans, S. Crothers, H. Eckardt and K. Pendergast, "Criticisms of the Einstein Field Equation" (CISP, Spring 2011).

[4] L. Felker, "The Evans Equations of Unified Field Theory" (Abramis, 2007).

[5] M. W. Evans, H. Eckardt and D. W. Lindstrom, plenary, Serbian Academy of Sciences and Arts, 2010.

[6] M .W. Evans and H. Eckardt, Contemporary Materials, 1(2), 112 –116 (2010).

[7] M .W. Evans, Contemporary Materials, 2(1), 1 –4 (2011).

[8] The ECE websites: www.aias.us, www.atomicprecision.com, www.et3m.net, www.upitec.org, archived at the National Library of Wales and the British Library on www.webarchive.org.uk

[9] K. Pendergast, "The Life of Myron Evans" (CISP, 2011).

[10] M. W. Evans and S. Kielich, "Modern Nonlinear Optics", (Wiley, 1992, 1993, 1997, 2001) in two editions and six volumes; M .W. Evans and L. B. Crowell, "Classical and Quantum Electrodynamics and the B(3) Field" (World Scientific, 2001); M. W. Evans and J.-P. Vigier, "The Enigmatic Photon" (Kluwer, 1994 –2002), in ten volumes softback and hardback; M. W. Evans and A. A. Hasanein, "The Photomagneton in Quantum Field Theory (World Scientific, 1994).

[11] S. P. Carroll, "Spacetime and Geometry: an Introduction to General Relativity" (Addison Wesley, New York, 2004).

[12] D. W. Lindstrom and H. Eckardt, J. Found. Phys. Chem., 1, 93 –102 (2011).

[13] D. W. Lindstrom and H. Eckardt, J. Found. Phys. Chem., 1, 103 –110 (2011).

[14] D. W. Lindstrom and H. Eckardt, J. Found. Phys. Chem., 1, 111 –116 (2011).

The post Einstein paradigm: refutations of Einsteinian relativity in spherically symmetric spacetimes

M.W. Evans[1*] and H. Eckardt[2]**

**Alpha Institute for Advanced Studies (AIAS) (www.aias.us); **Unified Physics Institute of Technology (UPITEC) (www.upitec.org)*

It is shown that any relativity theory based on the Schwarzschild type of infinitesimal line element is self contradictory in a spherically symmetric spacetime. This result is independent of any field equation. The Einstein / Schwarzschild general relativity is refuted conclusively by this new theorem. It follows that the well known Einsteinian theories are obsolete: perihelion precession, the deflection of light due to gravitation, the gravitational red shift, the theory of gravitational radiation, the theory of frame dragging, the theory of gravitational time delay, the theory of black holes, the cosmological red shift and the big bang theory of cosmology. In the post Einstein era the search has begun for a new relativity based on the ECE theory of unified physics.

Keywords: The post Einstein paradigm shift, refutations of Einsteinian general relativity, ECE theory of unified physics.

1 Introduction

In recent papers of this series [1–10] the standard Einsteinian theory of relativity has been refuted conclusively in spherically symmetric spacetimes defined by the so called 'Schwarzschild' metric (UFT190 ff and UFT200 on www.aias.us). In historical truth [11] Schwarzschild did not give this erroneous solution. The correct solution [11] does not have a singularity and does not predict such dogmatic ideas as 'black holes'. In the late nineteen fifties the Einsteinian general relativity was refuted completely by experimental data from galaxies, notably the velocity curve. It is therefore not possible to assert that the same theory has been tested with great precision with other data – from the solar system. This

[1]e-mail: EMyrone@aol.com
[2]e-mail: horsteck@aol.com

dogmatic adherence to an experimentally refuted theory characterized general relativity in the late twentieth century, and dogma is scientifically worthless. These refutations lead us into the post Einsteinian era characterized by a major paradigm shift of natural philosophy.

In Section 2 the refutation is given of a relativity theory in any Schwarzschild type spherically symmetric spacetime based on an infinitesimal line element. The methods of the Einstein theory itself are used to show straightforwardly that the theory is self contradictory and flawed irretrievably. It is plain wrong. This paradigm shift is therefore different in nature from those such as quantum mechanics, in which the original theory, classical mechanics, is considered to be correct within limits. Well known geometrical methods [12] are used to define the most general infinitesimal line element of the spherically symmetric spacetime, and the standard lagrangian methods of relativity applied in precisely the same way as in general relativity with the commonly called 'Schwarzschild' metric. Therefore, in Section 2 the standard methods of Einsteinian relativity are used to show that the Schwarzschild type theory is self contradictory and collapses in all spherically symmetric spacetimes or mathematical spaces of any dimension. This result is true irrespective of any constraint supplied by any field equation.

In Section 3 some absurd consequences of the 'Schwarzschild' metric are reviewed.

2 Refutation in the spherically symmetric spacetime

Consider the most general infinitesimal line element of spherically symmetric spacetime [12]:

$$ds^2 = c^2 d\tau^2 = m(r,t)c^2 dt^2 - n(r,t)df^2 - r^2 d\theta^2 \tag{1}$$

in cylindrical polar coordinates (r,θ). The theory is developed for the sake of clarity and simplicity in the plane defined by:

$$dZ^2 = 0 \tag{2}$$

but can be developed also in spherical polar coordinates and any system of coordinates in a mathematical space with any number of dimensions. Here τ is the proper time, c is the speed of light in vacuo, t the time in the observer frame, and in general $m\ (r,t)$ and $n(r,t)$ are any two functions of r and t. By definition:

$$d\mathbf{r} \cdot d\mathbf{r} = v^2 dt^2 = n(r,t)dr^2 + r^2 d\theta^2 \tag{3}$$

where the total velocity v in the observer frame is defined by:

$$v^2 = n(r,t)\left(\frac{dr}{dt}\right)^2 + r^2\left(\frac{d\theta}{dt}\right)^2. \tag{4}$$

Therefore:

$$ds^2 = c^2 d\tau^2 = m(r,t)c^2 dt^2 - d\mathbf{r}\cdot d\mathbf{r}. \tag{5}$$

The methods of Einsteinian relativity [12] are based on the definition of the lagrangian:

$$\mathcal{L} = \frac{1}{2}mc^2 = \frac{1}{2}\left(mm(r,t)c^2\left(\frac{dt}{d\tau}\right)^2 - mn(r,t)\left(\frac{dr}{dt}\right)^2 - mr^2\left(\frac{d\theta}{d\tau}\right)^2 \right) \tag{6}$$

where m is the mass of a particle or an object such as a planet in a planar orbit. The Euler Lagrange equations give two constants of motion as is well known, the total energy E and the total angular momentum L. These are conserved quantities because of the principles of conservation of total energy and conservation of total momentum. They are:

$$E = m(r,t)mc^2 \frac{dt}{d\tau} \tag{7}$$

and

$$L = mr^2 \frac{d\theta}{d\tau}. \tag{8}$$

From Eqs. (3) and (5):

$$\frac{dt}{d\tau} = \left(m(r,t) - \frac{v^2}{c^2} \right)^{-1/2}. \tag{9}$$

In special relativity:

$$m(r,t) \rightarrow 1 \tag{10}$$

so Eq. (9) reduces to the Lorentz factor.

Using the chain rule of differentiation:

$$\frac{d\theta}{d\tau}=\frac{d\theta}{dt}\frac{dt}{d\tau}=\frac{\omega dt}{d\tau} \tag{11}$$

It follows from Eqs. (7) and (8) that the angular velocity is:

$$\omega=\frac{d\theta}{dt}=\frac{cmb(r,t)}{r^2} \tag{12}$$

where

$$b=\frac{Lc}{E}. \tag{4}$$

From Eq. (1):

$$mc^2=mm(r,t)c^2\left(\frac{dt}{d\tau}\right)^2-mn(r,t)\left(\frac{dr}{d\tau}\right)^2-mr^2\left(\frac{d\theta}{d\tau}\right)^2=$$
$$=\frac{E^2}{(m,r)mc^2}-mn(r,t)\left(\frac{dr}{d\tau}\right)^2-\frac{L^2}{mr^2}. \tag{14}$$

Therefore:

$$mn(r,t)\left(\frac{dr}{d\tau}\right)^2=\frac{E^2}{m(r,t)mc^2}-mc^2-\frac{L^2}{mr^2}. \tag{15}$$

Apply to this equation the chain rule:

$$\frac{dr}{d\tau}=\frac{dr}{d\theta}\frac{d\theta}{d\tau}=\frac{L}{mr^2}\frac{dr}{d\theta} \tag{16}$$

to obtain:

$$\frac{mn(r,t)L^2}{m^2r^4}\left(\frac{dr}{d\theta}\right)^2=\frac{E^2}{m(r,t)mc^2}-mc^2-\frac{L}{mr^2}. \tag{17}$$

The orbital equation is therefore:

$$\left(\frac{dr}{d\theta}\right)^2 = \frac{mr^4}{n(r,t)L^2}\left(\frac{E}{m(r,t)mc^2} - mc^2 - \frac{L}{mr^2}\right) =$$
$$= \frac{r^4}{n(r,t)}\left(\frac{1}{m(r,t)b^2} - \left(\frac{1}{a^2} + \frac{1}{r^2}\right)\right) \tag{18}$$

where:

$$a = \frac{L}{mc},\; b = \frac{Lc}{E}. \tag{19}$$

Now use the chain rule again to find that:

$$\left(\frac{dr}{dt}\right)^2 = \omega^2\left(\frac{dr}{d\theta}\right)^2 = \left(\frac{cmb(r,t)}{r^2}\right)^2\left(\frac{dr}{d\theta}\right)^2 =$$
$$= c^2\frac{m(r,t)}{n(r,t)}\left(1 - b^2 m(r,t)\left(\frac{1}{a^2} + \frac{1}{r^2}\right)\right). \tag{20}$$

From Eq. (4) the total velocity v is:

$$v^2 = c^2 m(r,t)\left(1 - b^2 m(r,t)\left(\frac{1}{a^2} + \frac{1}{r^2}\right)\right) + \left(\frac{cbm(r,t)}{r^2}\right)^2 =$$
$$= c^2 m(r,t)\left(1 - \left(\frac{mc^2}{E}\right)^2 m(r,t)\right). \tag{21}$$

Note carefully that the total velocity does not depend on $n(r,t)$, and that Eq. (21) is the same result as obtained in UFT194 for the 'Schwarzschild' metric. From Eqs. (7), (8) and (9) an independent expression is obtained for $m(r,t)$ in terms of the total linear velocity:

$$m(r,t) = \frac{1}{2}\left(\frac{E}{mc^2}\right)\left(1 \pm \left(1 - \frac{4v^2}{c^2}\left(\frac{mc^2}{E}\right)^2\right)^{\frac{1}{2}}\right). \tag{22}$$

The Schwarzschild type line elemet is defined by:

$$m(r,t) = m(r) = 1 - \frac{r_0}{r} = n^{-1}(r). \tag{23}$$

In this case, the orbit is defined by:

$$\left(\frac{dr}{d\theta}\right)^2 = r^4\left(\frac{1}{b^2} - \left(1 - \frac{r}{r_0}\right)\left(\frac{1}{a^2} + \frac{1}{r^2}\right)\right). \tag{24}$$

This is claimed in the Einstein–Schwarzschild theory to be the orbit of a precessing ellipse. However, a precssing ellipse is defined by:

$$r = \frac{\alpha}{1 + \epsilon\cos(x\theta)} \tag{25}$$

where 2α is the right latitude, ϵ is the eccentricity and x is the precession constant. It follows from Eqs. (25) and (25) that:

$$\begin{aligned} \sin^2(x\theta) &= \left(\frac{\alpha}{x\epsilon}\right)^2\left(\frac{1}{b^2} - \left(1 - \frac{r}{r_0}\right)\left(\frac{1}{a^2} + \frac{1}{r^2}\right)\right) = \\ &= \left(\frac{\alpha}{x\epsilon}\right)^2\left(\frac{1}{b^2} - \frac{1}{a^2} + \frac{r_0}{a}\frac{1}{r} - \frac{1}{r^2} + \frac{r_0}{r^3}\right). \end{aligned} \tag{26}$$

However, from Eq.(25):

$$\sin^2(x\theta) = \left(1 - \frac{1}{\epsilon^2}\right) + \frac{2\alpha}{\epsilon^2}\frac{1}{r} - \left(\frac{\alpha}{\epsilon}\right)^2\frac{1}{r^2}. \tag{27}$$

Clearly, Eqs. (26) and (27) are not the same, thus refuting the Einstein–Schwarzschild theory.

The 'Schwarzschild' metric assumes that $m(r,t)$ varies with r as follows:

$$m(r,t) = m(r) = 1 - \frac{r_0}{r} \tag{28}$$

so is refuted directly. Hre r0 is the 'Schwarzschild' radius. The correct Schwarzschild metric of a letter of December 1915 from him to Einstein is in fact [11]:

$$m(r) = 1 - \frac{\gamma}{\left(r^3 + \alpha^3\right)^{1/3}} \tag{29}$$

References

[1] M.W. Evans, Ed., J. Found. Phys. Chem., six issues a year (Cambridge International Science Pub lishing, from June 2011, CISP, www.cisp-publishing.com)
[2] M. W. Evans, S. Crothers, H. Eckardt and K. Pendergast, "Criticisms of the Einstein Field Equation" (CISP, 2011).
[3] M W. Evans, H. Eckardt and D.W. Lindstrom, "Generally Covariant Unified Field Theory" (Abramis Academic, 2005 - 2011) in seven volumes.
[4] M W. Evans, H. Eckardt and D. W. Lindstrom, "ECE Theory of H Bonding", plenary lecture published in Proc. Int. Conf. Water, H Bonding, Nanomaterials and Nanomedicine, Serbian Academy, 2010.
[5] M.W. Evans and H. Eckardt, Contemporary Materials, 1(2), 112, (2010).
[6] M.W. Evans, Contemporary Materials, 2(1), 1 (2011).
[7] The www.aias.us website archived on www.webarchive.org.uk by the National Library of Wales and British Library; also www.upitec.org, www.atomicprecision.com, and www.et3m.net.
[8] L. Felker, "The Evans Equations of Unified Fiedl Theory" (Abramis 2007), translated into Spanish by Alex Hill on www.aias.us.
[9] K. Pendergast, "The Life of Myron Evans" (CISP, 2011).
[10] M.W. Evans, papers and books back to 1992 on the Omnia Opera Section of www.aias.us.
[11] A.A. Vankov, www.babin.net/eeuro.vankov.pdf
[12] S.P. Carroll, "Spacetime and Geometry: an Introduction to General Relativity" (Addison Wesley, New York, 2004 and open source notes).

The measurement of photon mass from the angle of deflection of one photon by the Sun

M.W. Evans[1*] and H. Eckardt[2]**

**Alpha Institute for Advanced Studies (AIAS) (www.aias.us); **Unified Physics Institute of Technology (UPITEC) (www.upitec.org)*

The mass of the photon is measured experimentally to be of the order of 10^{-57} kg using the experimental data that show that the orbit of a mass m around the sun is a conical section. The orbit of the photon of this mass is a hyperbola with a very large eccentricity, meaning that it is deflected very slightly as is well known. Einsteinian general relativity is not used, and in this paper is refuted conclusively in several ways, some of which are simple to understand, and which use simple algebra.

Keywords: ECE theory, mass of the photon, refutations of Einsteinian general relativity.

1. Introduction

It is shown in this paper that the measurement of the mass of the photon is straightforward if it is assumed that the photon mass m orbits the sun as any other mass M, in a conical section. The only thing that is known experimentally [1–10] is that light is deflected by the sun by a tiny amount, a few arc seconds only. It would not be known without other data whether the orbit is a closed precessing ellipse, or some other conical section such as a hyperbola, but it is known that the photon mass is less than about 10^{-52} kg, various estimates are available, so that its orbital eccentricity is very large. The orbit is therefore a hyperbola. In Section 2 it is shown that the photon mass is of the order of 10^{-57} kg if it is assumed that its orbit is a conical section. The Newtonian limit of a precessing hyperbola is a static hyperbola, and the tiny mass of the photon must mean that the eccentricity of this hyperbola is very large. The same is true for the right magnitude of the hyperbola. There is nothing in the Newtonian analysis to refute the assumption that the photon's orbit is a hyperbola. This conclusion is intuitively clear from the very small deflection of the photon by the sun.

[1]e-mail: EMyrone@aol.com

[2]e-mail: horsteck@aol.com

Therefore it is possible to derive the deflection angle and photon mass directly from the assumed orbit in the Newtonian limit. It is not necessary to use general relativity to find the mass of the photon, and indeed, in subsequent sections of this paper, it is shown that the Einsteinian general relativity is riddled with errors, and should no longer be used. It is thoroughly obsolete and has been replaced by the ECE theory of unified physics [1–10].

2. The mass of the photon

Consider an object of mass m orbiting an object of mass M in a precessing conical section [11]:

$$r = \frac{\alpha}{1+\epsilon\cos(x\theta)} \tag{1}$$

where 2α is the right magnitude, ϵ is the eccentricity, x is the precession constant, and where the coordinates in a plane are the cylindrical polar coordinates r and θ. It follows from Eq. (1) that:

$$\frac{dr}{d\theta} = \frac{x\epsilon}{\alpha} r^2 \sin(x\theta). \tag{2}$$

Therefore:

$$\frac{d\theta}{dr} = \frac{\alpha}{x\epsilon}\frac{1}{r}\left(r^2 - \frac{1}{\epsilon^2}(\alpha - r)^2\right)^{-1/2}. \tag{3}$$

If the distance of closest approach of the mass m to the mass M is R_o, then the object m is deflected by the angle:

$$\Delta\theta = \frac{2\alpha}{x\epsilon}\int_{R_o}^{\infty} \frac{dr}{r\left(r^2 - \frac{1}{\epsilon^2}(\alpha - r)^2\right)^{1/2}}. \tag{4}$$

This integral may be evaluated analytically and is:

$$\Delta\theta = \frac{2}{x}\left(\sin^{-1}\frac{1}{\epsilon} - \sin^{-1}\left(\frac{1}{\epsilon} - \frac{\alpha}{R_o}\right)\right). \tag{5}$$

If there is no deflection then:

$$\Delta\theta = 0 \tag{6}$$

and this result is equivalent from Eq. (5) to:

$$R_o \to \infty. \tag{7}$$

When the object m is infinitely distant from the object M it is not deflected and its trajectory is a straight line. In the limit:

$$x \to 1 \tag{8}$$

the conical section becomes static. In Newtonian theory [11] it is described by:

$$\alpha = \frac{L^2}{m^2 MG} \tag{9}$$

and

$$\epsilon = \left(1 + \frac{2EL^2}{m^3 M^2 G^2}\right)^{1/2} \tag{10}$$

where E is the total energy, L is the total angular momentum and G is Newton's constant. It is seen that m must be identically non-zero, the photon must have mass.

By definition the photon is the quantum of energy, so for one photon:

$$E = \hbar\omega, \; L = \hbar \tag{11}$$

where ω is the angular frequency. Consider therefore the orbit of one photon in the Newtonian limit defined by Eqs. (8), (9) and (10). Its angle of deflection is:

$$\Delta\theta = 2\left[\sin^{-1}\left(1 + \frac{2EL^2}{m^3 M^2 G^2}\right)^{-1/2} - \right. \\ \left. -\sin^{-1}\left(\left(1 + \frac{2EL^2}{m^3 M^2 G^3}\right)^{-1/2} - \frac{L^2}{m^2 MGR_o}\right)\right] \tag{12}$$

and this equation is also true for any object of mass m orbiting any object of mass M in the Newtonian limit. As in UFT150B (www.aias.us) the following values can be used:

$$\begin{aligned} \hbar &= 1.05459\times10^{-34}\ \text{Js} \\ R_o &= 6.955\times10^{8}\ \text{m} \\ M &= 1.9891\times10^{30}\text{kg} \\ G &= 6.67428\times10^{11}\text{m}^3\text{kg}^{-1}\text{s}^{-2} \end{aligned} \tag{13}$$

where R_o is the sun's radius and M is its mass. Equation (12) becomes:

$$\begin{aligned} \Delta\theta = 2\Bigg[& \sin^{-1}\left(1+\frac{1.33094\times10^{-170}}{m^3}\right)^{-1/2} - \\ & -\sin^{-1}\left(\left(1+\frac{1.33094\times10^{-170}}{m^3}\right)^{-1/2} - \frac{1.2045\times10^{-119}}{m^2}\right)\Bigg]. \end{aligned} \tag{14}$$

Assuming that the angular frequency of the photon is in the visible range:

$$\omega \sim 10^{16}\text{rad s}^{-1} \tag{15}$$

the measured deflection is:

$$\Delta\theta = (8.4848 \pm 0.003)\times10^{-6}\text{rad}. \tag{16}$$

Tables of elementary particle masses give the photon mass as less than about 10^{-52} kg, so the following Maclaurin series may be used:

$$\sin^{-1}x = x+\frac{x^3}{6}+\frac{3}{40}x^5+\ldots, |x|<1,\ \left|\sin^{-1}x\right|<\frac{\pi}{2}, \tag{17}$$

and the first term of this series is sufficient for visible frequency radiation. So Eq. (14) reduces to:

$$\Delta\theta = \frac{2.409\times10^{-119}}{m^2} \tag{18}$$

giving the following mass of the photon:

$$m = 1.685 \times 10^{-57}\ \text{kg}. \tag{19}$$

Using this value back in Eq. (14), it is seen that approximation:

$$\sin^{-1}x \sim x \tag{20}$$

is valid because:

$$\left(1+\frac{1.33094\times10^{-170}}{1.685^{3}\times10^{-171}}\right)^{-1/2}=0.5142 \tag{21}$$

and

$$\frac{1.2045\times10^{-119}}{1.685^{2}\times10^{-114}}=4.242\times10^{-6}. \tag{22}$$

The value (19) is an order of magnitude, but it is the first time that the mass of the photon has been determined, a hundred year old aim of physics. This estimate agrees well with other estimates that give the mass as less than 10^{-52} kg. This mass means that the eccentricity of the photon's hyperbolic orbit is:

$$\epsilon = 1.945 \tag{23}$$

and its half right magnitude is:

$$d = 2.950 \text{ m} \tag{24}$$

Note that this value of the photon mass does not depend on the frequency in the approximations used, but in general the photon mass depends on frequency and cannot be an elementary particle mass. As in UFT158 ff. It is developed in terms of R theory.

This determination of photon mass refutes the massless photon theory of the U(1) sector of the standard model, and refutes the Higgs boson theory. The existence of photon mass corroborates the B(3) theory [1–10].

3. Definitive and simple refutation of Einsteinian general relativity

It is very simple to refute Einstein's general relativity (EGR) and this fact has been known for nearly a century. This is why general relativity cannot determine the photon mass from the deflection of light by gravitation. In these sections a series of refutations is given. The EGR theory for a planar orbit is based on the infinitesimal line element:

$$ds^2 = c^2 d\tau^2 = \left(1-\frac{r_0}{r}\right)c^2\,dt^2 - \left(1-\frac{r_0}{r}\right)^{-1} dr^2 - r^2 d\theta^2 \tag{25}$$

in the plane:

$$dZ^2 = 0 \tag{26}$$

where the distance r_0 is defined by:

$$r_0 = \frac{2MG}{c^2}. \tag{27}$$

The EGR theory claims incorrectly that this line element produces a precessing ellipse. It is claimed that [1–10, 12]:

$$\left(\frac{x\epsilon}{\alpha}\right)^2 \sin^2(x\theta) = ?\ \frac{1}{b^2} - \left(1 - \frac{r_0}{r}\right)\left(\frac{1}{a^2} + \frac{1}{r^2}\right) \tag{28}$$

where the constants a and b are:

$$a = \frac{L}{mc},\ b = \frac{cL}{E}. \tag{29}$$

If this claim were true then:

$$\sin^2(x\theta) = ?\ \left(\frac{\alpha}{x\epsilon}\right)^2 \left(\frac{1}{b^2} - \frac{1}{a^2} + \frac{r_0}{a}\frac{1}{r} - \frac{1}{r^2} + \frac{r_0}{r^3}\right) \tag{30}$$

However, from Section 2:

$$\sin^2(x\theta) = \left(1 - \frac{1}{\epsilon^2}\right) + \frac{2\alpha}{\epsilon}\frac{1}{r} - \left(\frac{\alpha}{\epsilon}\right)^2 \frac{1}{r^2} \tag{31}$$

and this function is derived directly from the observations in astronomy of a precessing ellipse or a conical section in general. This function is not Eq. (30), so EGR is refuted, QED.

4. Refutations based on light deflection due to gravitation

In Section 2 it was shown that light deflection due to gravitation depends on m in the Newtonian limit. In EGR it is claimed erroneously that the Newtonian limit is

$$\Delta\theta = ?\ \frac{2GM}{c^2 R_o} \tag{32}$$

but this limit does not depend on m. It cannot be correct, QED. EGR claims that the light deflection is:

$$\Delta\theta = ?\ \frac{4GM}{c^2 R_o} \tag{33}$$

but again this does not depend on the mass m. It cannot therefore reduce to the correct Newtonian limit given in Section 2. In UFT 150B the various errors made by Einstein in his light deflection calculation are discussed. It is clear that this well known calculation is incorrect because it starts from Eq. (25), and Eq. (25) as shown in Section 3 does not produce the conic section from which the deflection must be calculated for elementary self consistency.

Einstein appears to have carried out this calculation in ref. (13), in a textbook such as that by Wald [14], the integral used by Einstein is described as:

$$\Delta\theta = 2\int_0^{1/R_o}\left(R_o^{-2} - 2MR_o^{-3} - u^2 + 2Mu^3\right)^{-1/2} du \tag{34}$$

where M is mass in reduced units. In S.I. Units it is MG/c^2. The method used to evaluate Eq. (34) is very obscure, it relies on:

$$\left.\frac{\partial(\Delta\theta)}{\partial M}\right|_{M=0} = ?\ \frac{1}{4b}. \tag{35}$$

However, M is a constant, and Eq. (35) is not an operation of elementary differentiation. In addition, Section 3 shows that this integral is incorrect.

All theories based on EGR are easily refuted: light deflection by gravitation, gravitational time delay, de Sitter precession, gravitational time delay, perihelion precession and gravitational radiation.

5. Numerous refutations based on geometry

Equation (25) is obtained after a tortuous series of incorrect assumptions about the general line element:

$$ds^2 = c^2 d\tau^2 = m(r,t)c^2 dt^2 - n(r,t)dr^2 - r^2 d\theta^2 \tag{36}$$

of spherical spacetime, where m and n are general functions of r and of time t [12]. The most basic and pervasive blunder of EGR is its assumption of a symmetric Christoffel connection, as discussed in previous work [1–10]. The antisymmetry of the connection is easily shown by consideration of the basic equation:

$$\left[D_\mu, D_\nu\right]V^\sigma = -\left(\Gamma^\lambda_{\mu\nu} - \Gamma^\lambda_{\nu\mu}\right)D_\lambda V^\rho + R^\rho_{\sigma_{\mu\nu}}V^\sigma \tag{37}$$

which is true in any space of any dimension. The commutator of covariant derivatives on the left hand side acts on a vector (or any tensor of any rank) to isolate the connection as shown on the right hand side. Here the Riemann tensor is well known to be the antisymmetric object:

$$R^\rho_{\sigma_{\mu\nu}} = -R^\rho_{\sigma_{\mu\nu}} = \partial_\mu\Gamma^\rho_{\nu\sigma} - \partial_\nu\Gamma^\rho_{\mu\sigma} + \Gamma^\rho_{\mu\lambda}\Gamma^\lambda_{\nu\sigma} - \Gamma^\rho_{\nu\lambda}\Gamma^\lambda_{\mu\sigma}. \tag{38}$$

It has the same commutator indices μ and ~ as the connection, so the latter is antisymmetric, QED:

$$R^{\rho}_{\sigma_{\mu\nu}} = -R^{\rho}_{\sigma_{\mu\nu}},\ \Gamma^{\lambda}_{\mu\nu} = -\Gamma^{\lambda}_{\nu\mu}. \tag{39}$$

If the indices were the same, the commutator, connection and curvature would all vanish. In this case the covariant derivate D_μ reduces to the ordinary derivative ∂_μ, and:

$$\left[\partial_\mu, \partial_\nu\right] V^\sigma = 0. \tag{40}$$

The torsion is well known [1–10, 12] to be:

$$T^{\lambda}_{\mu\nu} = \Gamma^{\lambda}_{\mu\nu} - \Gamma^{\lambda}_{\nu\mu} \tag{41}$$

and is always non-zero. The basic meaningless of EGR is that it is based on a self contradictory idea that the Riemman tensor is antisymmetric but that the connection is symmetric. Obviously, both are antisymmetric. It cannot be claimed that the Riemman tensor is symmetric, it cannot be claimed that the connection is symmetric, it cannot be claimed that the commutator is symmetric. There is no symmetric part to any of these objects because they are generated by a round trip in a space with curvature and torsion, and thus by an intrinsically antisymmetric commutator. If any such objects were symmetric there would be no curvature and no torsion.

In consequence of this blunder the following incorrect equation is used in EGR:

$$\Gamma^{\sigma}_{\mu\nu} = ?\ \frac{1}{2} g^{\sigma\rho}\left(\partial_\mu g_{\nu\rho} + \partial_\nu g_{\rho\mu} - \partial_\rho g_{\mu\nu}\right) \tag{42}$$

where $g_{\mu\nu}$ is the metric. This equation is no longer true for an antisymmetric connection. Similarly, the Ricci tensor:

$$R_{\sigma\mu} = R^{\lambda}_{\sigma\lambda\mu} \tag{43}$$

is obtained in EGR from an incorrectly evaluated Riemman tensor, a Riemann tensor obtained from the incorrect (42). The second Bianchi idenity used in EGR is incorrect for the same reason, and the Einstein field equation is incorrect because it is based on the second Bianchi identity. The next in a series of blunders in EGR is to assume that the Ricci tensor vanishes. Crothers has shown that this is not true (www.aias.us). In order to derive Eq. (25) from Eq. (36), the latter is written as [12]:

$$ds^2 = c^2 dt^2 e^{2\alpha(r,t)} - e^{2\beta(r,t)} dr^2 - r^2 d\theta^2. \tag{44}$$

The incorrect assumption of a null Ricci tensor produces the incorrect results:

$$\beta = \beta(r) \tag{45}$$

$$\alpha = f(r) + g(t) \tag{46}$$

The function $g(t)$ is 'eliminated' using a dubious procedure [12]:

$$dt \rightarrow \exp(-g(t))dt \tag{47}$$

to give:

$$ds^2 = e^{2\alpha(r)}c^2dt^2 - e^{2\beta(r)}dr^2 - r^2d\theta^2. \tag{48}$$

The incorrect use of the null Ricci tensor produces the incorrect result:

$$\alpha = -\beta + \text{constant} \tag{49}$$

in which it is asserted without proof that:

$$\alpha = -\beta. \tag{50}$$

This is known as 'scaling coordinates', but it is entirely arbitrary. This arbitrary procedure leads to:

$$e^{2\alpha}\left(2r\partial_1 \alpha + 1\right) = 1 \tag{51}$$

i.e.:

$$\partial_1\left(re^{2\alpha}\right) = 1 \tag{52}$$

which means that:

$$e^{2\alpha} = 1 + \frac{\mu}{r}. \tag{53}$$

The Einstein field equation is used only in its weak field limit to claim that:

$$\mu = \frac{2GM}{c^2} \tag{54}$$

but this procedure is riddled with errors and so the line element (25) is entirely without any meaning.

If this were not enough, EGR's light deflection and black hole theory are based on the idea of a null geodesic (see UFT150B on www.aias.us):

$$ds^2 = ?\ 0 \tag{55}$$

However, Eq. (25) is:

$$c^2 = c^2\left(\frac{dt}{d\tau}\right)^2\left(1-\frac{r_0}{r}\right)-\left(1-\frac{r_0}{r}\right)^{-1}\left(\frac{dr}{d\tau}\right)^2 - r^2\left(\frac{d\theta}{d\tau}\right)^2 \tag{56}$$

and is obviously incompatible with Eq. (55) because Eq. (55) makes Eq. (56) singular. The entire history of black hole theory can be rejected, it is a meaningless theory. Black hole theory proceeds with the meaningless mathematical assumption [12]:

$$ds^2 = ?\ 0 = \left(1-\frac{2GM}{c^2 r}\right)c^2dt^2 - \left(1-\frac{2GM}{c^2 r}\right)^{-1} dr^2. \tag{57}$$

This assumption is made yet more obscure by a tortuous choice of coordinates:

$$ct = ? \pm\left(r + \frac{2GM}{c^2}\log_e\left(\frac{rc^2}{2GM}-1\right)\right) + \text{constant} \tag{58}$$

a mathematical contrivance used to redefine the line element as

$$ds^2 = ?\left(1-\frac{2GM}{c^2 r}\right)\left(c^2dt^2 - dr^{*2}\right) - r^2 d\theta^2. \tag{59}$$

There are very many sequential errors, the most pervasive one is to define the 'event horizon' as:

$$r = ?\frac{2GM}{c^2 r}. \tag{60}$$

All black hole metrics were thoroughly refuted with computer algebra in ref. (1).

References

[1] M.W. Evans, S. Crothers, H. Eckardt and K. Pendergast, "Criticisms of the Einstein Field Equation" (Cambridge International Science Publishing, CISP, www.cisp-publishing.com, Spring 2011).

[2] M.W. Evans, Ed. J. Foundations of Physics and Chemistry, (CISP, June 2011 onwards), six editions a year.

[3] M.W. Evans, H. Eckardt and D. W. Lindstrom, "Generally Covariant Unified Field Theory" (Abramis Academic, 2005 to 2011) in seven volumes.

[4] L. Felker, "The Evans Equations of Unified Field Theory" (Abramis 2007, Spanish translation by Alex Hill on www.aias.us).

[5] K. Pendergast, "The Life of Myron Evans" (CISP, Spring 2011).

[6] The ECE websites: www.webarchive.org.uk, www.aias.us, www.atomicprecision.com, www.e3tm.net, www.upitec.org.

[7] M.W. Evans and S. Kielich (Eds.), "Modern Nonlinear Optics" (Wiley, 1992, 1993, 1997, 2001), two editions and six volumes.
[8] M.W. Evans and L. B. Crowell, "Classical and Quantum Electrodynamics and the B(3) Field" (World Scientific, 2001).
[9] M.W. Evans and J.-P. Vigier, "The Enigmatic Photon" (Kluwer 1994 to 2002), in ten volumes hardback and softback.
[10] M.W. Evans and A. A. Hasanein, "The Photomagneton in Quantum Field Theory" (World Scientific, 1994).
[11] J.B. Marion and S. T. Thornton, "Classical Dynamics" (HBC, New York 1988, 3rd edition).
[12] S.P. Carroll, "Spacetime and Geometry: an Introduction to General Relativity" (Addison Wesley, New York, 2004).
[13] A. Einstein, Proceedings of the Royal Prussian Academy, November 1915.
[14] R.M. Wald, "General Relativity" (Chicago University Press, 1984).